Experiments with Digital Electronics
50 practical circuits

Experiments with Digital Electronics
50 practical circuits

Burkhard Kainka / Lars Gollub

Elektor International Media
www.elektor.com

All rights reserved. No part of this book may be reproduced in any material form, including photocopying, or storing in any medium by electronic means and whether or not transiently or incidentally to some other use of this publication, without the written permission of the copyright holder except in accordance with the provisions of the Copyright, Designs and Patents Act 1988 or under the terms of a licence issued by the Copyright Licensing Agency Ltd, 90 Tottenham Court Road, London, England W1P 9HE. Applications for the copyright holder's written permission to reproduce any part of this publication should be addressed to the publishers.

The publishers have used their best efforts in ensuring the correctness of the information contained in this book. They do not assume, and hereby disclaim, any liability to any party for any loss or damage caused by errors or omissions in this book, whether such errors or omissions result from negligence, accident or any other cause.

British Library Cataloguing in Publication Data
A catalogue record for this book is available from the British Library

ISBN 978-0-905705-97-2

Translation (from German): Brian Tristam Williams (SA)
Prepress production: Eric A.J. Bogers, Aschendorf (DE)
Cover Design: Etcetera, Aachen (DE)
First published in the United Kingdom 2010
Printed in the Netherlands by Wilco, Amersfoort

099004-UK

Table of contents

	Foreword	9
1	**Introduction**	**11**
	1.1 Components	11
2	**Digital Fundamentals**	**17**
	2.1 Bits and binary numbers	17
	2.2 The basic digital functions	17
	2.3 IC families	19
3	**Gate Functions**	**21**
	3.1 NAND gate	23
	3.2 NAND gate as inverter	26
	3.3 AND gate	31
	3.4 NOR gate	33
	3.5 NOR gate as inverter	35
	3.6 OR gate	39
	3.7 NOR becomes AND	41
	3.8 NOR becomes NAND	43
	3.9 NAND becomes OR	45
	3.10 NAND becomes NOR	47
	3.11 XOR	49
	3.12 XNOR	51
	3.13 The majority function	53
4	**Flip-Flops**	**55**
	4.1 RS flip-flop from NOR gates	55
	4.2 RS flip-flop from NAND gates	58
	4.3 JK flip-flop as RS flip-flop	60
	4.4 The JK flip-flop	62

TABLE OF CONTENTS

	4.5	Shift registers	65
	4.6	Circular shift registers	70
5	**Counters**		**73**
	5.1	Counter to 3	73
	5.2	4-bit counter	76
	5.3	Synchronous counter	79
	5.4	Up/down counter	82
6	**The Digit Display**		**87**
	6.1	Digit segments	87
	6.2	Seven-segment decoder	89
	6.3	0 to 9 counter	91
	6.4	9 to 0 countdown	93
7	**Oscillators**		**95**
	7.1	Blinking light	95
	7.2	Flip-flop blinker	97
	7.3	Metronome	99
	7.4	Tone generator	101
8	**Applications**		**103**
	8.1	Light-controlled tone	103
	8.2	Mini organ	105
	8.3	Siren	107
	8.4	Twilight switch	109
	8.5	Alarm system	111
	8.6	Light-activated alarm system	113
	8.7	Running light	115
	8.8	Traffic light control	119
	8.9	Turn-off delay	122
	8.10	Turn-on delay	124
	8.11	Time switch	126
	8.12	Hallway light timer	128
	8.13	Simple random generator	131
	8.14	Digital roulette	133
	8.15	Digital dice	135

9	**SGS Datasheets**	**139**
	HCF4001B	139
	HCF4027B	147
	HCF4093B	155
	HCF4511B	161
	Index	175

Foreword

Digital electronics are central to modern technology. Take a look inside a modern appliance, and you'll find highly integrated circuits, microcontrollers, programmable logic devices, etc. Simple digital ICs such as the 4001 and the 4093, however, are rarely found. You might assume that these devices are obsolete, and that dealing with them is redundant. Should we not begin immediately with microcontrollers, since everything we need for digital electronics is, after all, right in there? We've decided to focus on relatively simple circuits consisting of gates, flip-flops and counters from the CMOS 4000 series. These are the fundamentals of digital electronics. Only one who really understands how a digital circuit works can successfully move on to more highly integrated devices. A quick look at any microcontroller datasheet makes it clear that the manufacturers of these devices assume a basic knowledge of digital electronics on the part of the reader – the datasheets are full of detailed circuit diagrams with individual digital building blocks. Even programming a microcontroller requires a knowledge of basic logical functions and relationships.

There are countless books dealing with the required theoretical knowledge, but a study of these fundamentals is best achieved using practical experiments. Building digital circuits is both fun and educational, and many of the 50 circuits presented here have practical applications. Someone with a general overview of the field will be well equipped to find the simplest solution for any application. Our overview uses digital CMOS ICs, which are introduced using practical experiments. Working on this book was a shared task: Lars Gollub developed all of the experiments, circuit diagrams and construction illustrations, while Burkhard Kainka wrote the text.

Have fun with the experiments!

Burkhard Kainka and Lars Gollub

1 Introduction

This book presents the basics of digital electronics with the help of digital devices from the CMOS 4000 series. All of the experiments are carried out on a solderless breadboard.

Anyone who works with control systems such as PLCs or microcontrollers has to know the basics. How data is processed, how output elements such as LEDs, displays and speakers are controlled, what the basic logical functions are, how registers and counters work, how inputs and outputs work – all areas that we'll examine using practical experiments.

The book is intended mainly for electronics hobbyists, students and trainees who need a basic introduction to digital electronics. This knowledge is also the foundation for larger-scale projects in microcontrollers and programming.

All experiments in this book are designed for self-study or individual training. They're also suitable for use in education, on-the-job training or for practical work at universities. Elektor does offer a component kit consisting of a large breadboard, ICs and further components (www.elektor.com/digitalexperiments), but naturally you may use components available on-hand from the junk box or build the circuits somewhat differently, e.g. on stripboard. However, construction on a breadboard has proven practical, as it allows quick and trouble-free modifications to be made. Often, it allows further variations upon the circuit to be easily developed and tested.

1.1 Components

A small set of components is used to carry out many experiments at a low cost. This makes the purchase of classroom sets for use in schools

1 INTRODUCTION

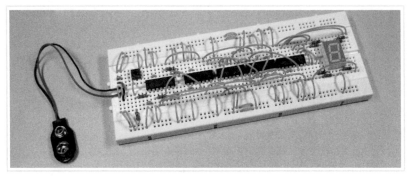

Figure 1.1   Typical experiment construction

easier. Also, all components are also available in a kit from Elektor (*www.elektor.com/digitalexperiments*). The resistors employed have four rings and an additional brown ring to indicate their 1 % tolerance.

List of components

Resistors:
3 resistors, 820 Ω
  (green-red-black-black)
8 resistors, 1 kΩ
  (brown-black-black-brown)
1 resistor, 3.3 kΩ
  (orange-orange-black-brown)
4 resistors, 10 kΩ
  (brown-black-black-red)
1 resistor, 100 kΩ
  (brown-black-black-orange)
2 resistors, 220 kΩ
  (red-red-black-orange)
1 light-dependent resistor (LDR)

Capacitors:
4 ceramic capacitors, 100 nF
  ("104" identifier)
1 electrolytic capacitor, 10 μF/16 V

Semiconductors:
7 red LEDs
1 yellow LED
1 green LED
1 seven-segment display with common cathode
1 diode, 1N4148

1 IC, 4001
  (quad NOR gate)
2 ICs, 4093
  (quad NAND gate with Schmitt-trigger inputs)
2 ICs, 4027
  (dual JK flip-flop)
1 IC, 4511
  (seven-segment decoder)

Miscellaneous:
1 breadboard, 640x200 contacts
2 m hook-up wire
4 switches
1 piezoelectric transducer
1 battery clip

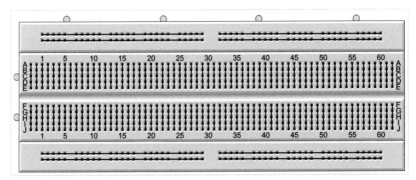

Figure 1.2   Breadboard terminal connections

The breadboard is suitable for discrete components and ICs in DIL packages. Figure 1.2 shows the terminal connections. Note that the four long power rails are interrupted in the middle of the board.

Each experiment has both a circuit diagram and a layout diagram. The layout diagrams are to give an overview of optimal, exact-fitting, carefully-angled wires. However, it is probably useful to use somewhat longer wires, which may be bent into the individual contacts. The wires may then easily easily reused in the experiments that follow. First, cut the individual wires with a side cutter. Separate the insulation at the ends for a length of about 6 mm. It may be effective first to cut the insulation around the wire with a sharp knife. Be careful not to cut into the actual conductor, as it may then break off easily in the contacts. With some practice, you can roll the wire between the knife and a tabletop, until the insulation is almost cut right through. It will then be easy to remove.

Sometimes, the wire can only be inserted using force. It may help to widen the contact with a pin. Also, if you cut the wire diagonally, a sharp end is created, which may be easier to insert.

The ICs used are 14- and 16-pin devices from the CMOS series. The images on the next page show the pin configurations. Usually, the power connections are in two corners, on Pin 7 (or 8) (ground) and Pin 14 (or 16) (positive supply voltage). In all of our experiments, ground is the same as the battery's negative terminal, even though the general

1 INTRODUCTION

symbol for ground will be used for illustration. The GND symbol is also common. The connection for the positive supply voltage is often indicated as VCC, which stands for common-collector voltage, although no bipolar transistors are used here. For this reason, the negative connection is shown in some datasheets as VEE (for emitter). Other common symbols are VSS (source) for ground, and VDD (drain) for the positive connection.

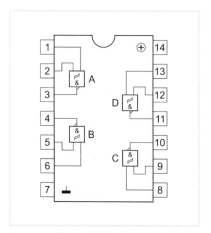

Figure 1.3   4093 pin layout

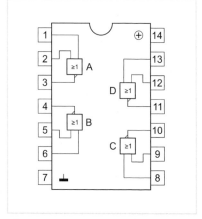

Figure 1.4   4001 pin layout

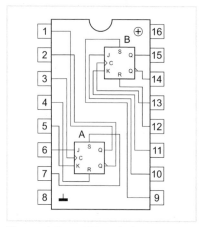

Figure 1.5   4027 pin layout

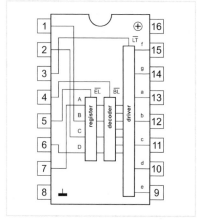

Figure 1.6   4511 pin layout

The ICs usually contain several identical function blocks, which are indicated by the letters A, B, C and D. In contrast with the datasheet, we indicate the JK flip-flop closest to the lower pin numbers as "A".

2 Digital Fundamentals

Digital means discrete and describes an environment with distinct states. In the simplest case of a single switch, there are only two states: On or off, yes or no, '1' or '0', true or false. We also speak of binary states, as a switch has exactly two states. A simple switch works digitally; it is either on or off, and there is nothing between these two states.

2.1 Bits and binary numbers

The smallest unit of information in digital technology is a bit, which is capable only of these two states, '1' and '0.' Should we have, for example, two switches, then each would represent one bit. In total, there would be four possible states between the two switches. With three bits, there are already eight states, and with four bits, 16 states can represent numbers between 0 and 15. A 4-bit binary number consists of four zeroes or ones and we can represent this as in Table 2.1. '0000' has a value of 0, '1100' is 12 and '1111' is 15. So, when considering four switches, we have a total of 16 possible states. Four units of digital information (bits) allow 16 individual states to be represented. By contrast, there is analogue technology. An analogue voltmeter can display any number of different voltages.

2.2 The basic digital functions

Digital circuits may, in principle, be built from switches. Even when there are as few as two switches, there are several wiring possibilities, each quite dependent on the application.

An example: There is a machine in a factory, which is operated using two large switches. A worker must hold both switches down in order to ensure that neither of his hands is within the danger range. The wiring is simple: The two switches are connected in series. The machine

2 DIGITAL FUNDAMENTALS

Bit 3 2^3 = 8	Bit 2 2^2 = 4	Bit 1 2^1 = 2	Bit 0 2^0 = 1	Value (decimal)
0	0	0	0	0+0+0+0=0
0	0	0	1	0+0+0+1=1
0	0	1	0	0+0+2+0=2
0	0	1	1	0+0+2+1=3
0	1	0	0	0+4+0+0=4
0	1	0	1	0+4+0+1=5
0	1	1	0	0+4+2+0=6
0	1	1	1	0+4+2+1=7
1	0	0	0	8+0+0+0=8
1	0	0	1	8+0+0+1=9
1	0	1	0	8+0+2+0=10
1	0	1	1	8+0+2+1=11
1	1	0	0	8+4+0+0=12
1	1	0	1	8+4+0+1=13
1	1	1	0	8+4+2+0=14
1	1	1	1	8+4+2+1=15

Table 2.1   Counting in binary

is turned on only when both Switch 1 and Switch 2 are closed. This is an example of the digital AND function.

All of the possible states may be represented in a so-called "truth table." The name comes from the fact that one may determine under which circumstances the output of the function is true ('1') or false ('0').

Input 1	Input 2	Output
0	0	0
0	1	0
1	0	0
1	1	1

Table 2.2   The AND function

OR is another important digital function, in which the two switches are connected in parallel. In this manner, we might, for example, connect an alarm siren. When either Switch 1 or Switch 2 is actuated or both are actuated together, the alarm sounds.

Input 1	Input 2	Output
0	0	0
0	1	1
1	0	1
1	1	1

Table 2.3   The OR function

The third basic function is NOT. Imagine an off switch – for example, a large, red emergency switch, fitted quite visibly on a machine. In the simplest case, the switch would open the circuit when activated. When we press the switch, the machine turns off.

Input	Output
0	1
1	0

Table 2.4   The NOT function

From these three basic functions, AND, OR and NOT, many more can be created. Eventually, switches wouldn't be sufficient, and one would use relays, for example, which are electrically-activated switches. An electromagnet usually operates several switch contacts simultaneously, which offers us several possibilities. Entire computers were once built using relays. Later, other active switching elements, such as valves and transistors, were used. The first digital devices were really large and complex. When many components were combined into integrated circuits, the digital electronics environment got simpler and cheaper.

2.3   IC families

Among the first digital ICs were the 7400-Series TTL devices (Transistor-Transistor Logic with bipolar transistors). They were often used in demonstration systems, e.g. for schools and training. The requirement of a stable, regulated 5-V power supply was disadvantageous.

Later, the 4000-Series CMOS ICs arrived. CMOS stands for Complimentary Metal-Oxide Semiconductor transistors, or n-channel and p-channel field-effect transistors. The complementary FETs are usually of the

2 DIGITAL FUNDAMENTALS

push-pull variety, wherein only one of the two conducts, taking the output to either GND or VCC. These ICs work on voltages of between 3 V and 15 V and are relatively slow, with switching speeds of a maximum of a few MHz. Newer ICs from the HCMOS-Series are substantially faster, and can thus be put to more demanding use.

We have opted for the 4000-Series CMOS ICs, as they make few demands on operating voltage and circuit layout. This enables the use of a 9-V battery and construction on a simple solderless breadboard. The wire lengths are not critical. By contrast, very fast digital ICs always require a board with well-planned ground planes and interconnections that are as short as possible, lest one have to deal with unwelcome surprises. Standard CMOS ICs are very well natured and undemanding and are ideal for training purposes.

The CMOS Series has many practical applications. When you need to realise a simple digital application in a cost-effective way, this series is your first choice. The freedom of operating voltage and extremely low current consumption speak for themselves. In many cases, one may even eliminate the power switch altogether and allow the circuit to operate on the battery voltage continuously.

3 Gate Functions

A basic logical circuit is known as a gate, because we can think of it as a gate that is either open or closed. That is, a circuit that turns the current on or off, allowing a voltage to reach the output, or not, depending on the input states. The basic gate types are:

AND: The schematic symbol has its inputs on the left and output on the right. The function is represented by the '&' symbol.

OR: The schematic symbol contains a greater-than-or-equal-to sign.

NOT (inverter): The schematic symbol contains a '1' and has a diagonal line to indicate the inverter at its output.

An AND gate requires five connections: Two inputs, an output, and two connections for the supply voltage. Standard ICs usually contain several gates instead of just one. For four AND gates, we need 14 connections, and the 4081 (quad AND) provides exactly this. Similarly, there is a quad OR gate, the 4071, which also has 14 legs.

For the NOT function, we can use the 4069 hex inverter. Because only a single input and output is required per gate, exactly six NOT gates fit within a 14-pin IC. From these three ICs, every possible digital circuit may be built.

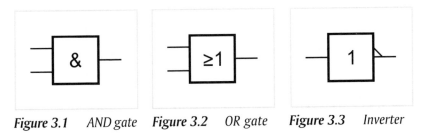

Figure 3.1   AND gate   *Figure 3.2*   OR gate   *Figure 3.3*   Inverter

3 GATE FUNCTIONS

AND and OR gates are seldom found in practical circuits. More often, we find NAND gates – a combination of an AND gate and an inverter. At the output of its symbol, an additional inversion line is drawn. NAND gates are far more universally applicable. If necessary, a NAND gate may be used as a simple inverter, and an OR gate may even be fashioned from three NAND gates. Sometimes, one would rather use a quad NAND gate instead of two ICs with different functions, even if we don't use all of the gates in the package. The goal is to have as few ICs as possible in a circuit. More universal ICs are also cheaper and easier to obtain.

Here, a special variant of the NAND gate is used – the 4093 NAND gate with Schmitt trigger inputs. A Schmitt trigger can detect a rising or falling analogue voltage at its inputs. At a 9 V operating voltage, the input voltage must reach approximately 6 V before it is considered to be in the '1' state. Thereafter, it needs to fall to 3 V or below, before it is considered to have returned to the '0' state. Between these switching points, there is a range (the hysteresis) in which the existing state will not change. This property is indicated with the additional symbol of two diagonal lines with a space, representing the hysteresis, between them. Therefore, there is always a unique input condition. This is especially important for switch de-bouncing and for oscillator circuits, which we will discuss below.

In addition to the NAND gate, a NOR gate is also frequently used. It is made up of an OR gate and an inverter.

Our first experiments are concerned with the basic gate functions and with the development of new functions using a combination of several gates.

Figure 3.4   *A NAND gate with Schmitt trigger inputs*   *Figure 3.5*   *A NOR gate*

3.1 NAND gate

The first experiment demonstrates the function of a NAND gate. It uses just one of the 4093's four NAND gates. The other three unused gates may not, however, be simply left alone, as CMOS inputs may assume random states, which, in extreme cases, may lead to higher current consumption or to circuit oscillation. The general rule is: Open CMOS inputs must be terminated. This is done by connecting them to ground. On the other hand, the outputs must be left unconnected.

The inputs in use must also have a definite input level. For this, so-called "pull-down" 10 kΩ resistors are used. With an open input, the input voltage will be 0 V (logic '0'). Press a switch, and the input voltage will jump to 9 V, accordingly (logic '1').

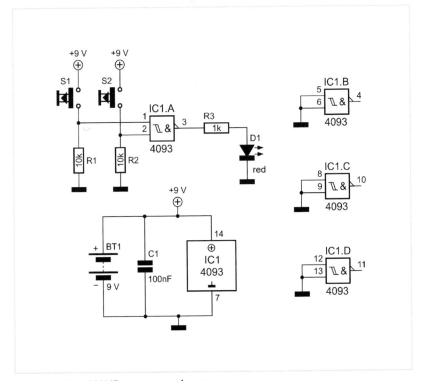

Figure 3.6   *NAND gate experiment*

In its quiescent state, the gate output has a voltage of either 0 V or 9 V. However, because the connected LED has a forward voltage of approximately 1.8 V, a resistor is necessary. Due to the finite internal resistance of the CMOS output, the output voltage drops somewhat when under load. With a 1 kΩ resistor, one can achieve a voltage drop of approximately 7 V. The LED draws a current of around 7 mA. Because a standard LED may operate on up to 20 mA, the current doesn't near the maximum value, but the LED is still relatively bright.

In this experiment, the LED is an indicator of the output state of the circuit: The LED on indicates a logic '1', and when it's off, this indicates a logic '0.'

All of the circuits in this book use the same schematic for the supply voltage. At the top and bottom edges of the experiment board are two supply rails for +9 V (upper edge) and ground (GND, lower edge). The rails are broken in the middle of the board, and must be connected together using four short wire bridges. In addition, to minimise electrical interference, a ceramic capacitor is connected between +9 V and ground. This wiring and the connection of the battery clip is the same for all experiments, and need not be modified. Each IC is then supplied with the necessary operating voltage by connecting it to the supply rail.

The battery clip has a red input wire (top) for the positive pole and a black wire (bottom) for the negative pole. Because the tinned wire braids are delicate, they should be inserted into the breadboard only once, so that the wires don't break off. For this reason, two small wire bridges are arranged so that they create a bridge and simultaneously serve as a strain relief for the battery wire.

> **When you build new circuits, always disconnect the battery from the clip, but leave the wire connected to the board.**

Test all four possible input state combinations and observe the output state. The behaviour of the NAND gate is illustrated in the truth table (Table 3.1).

NAND GATE 3.1

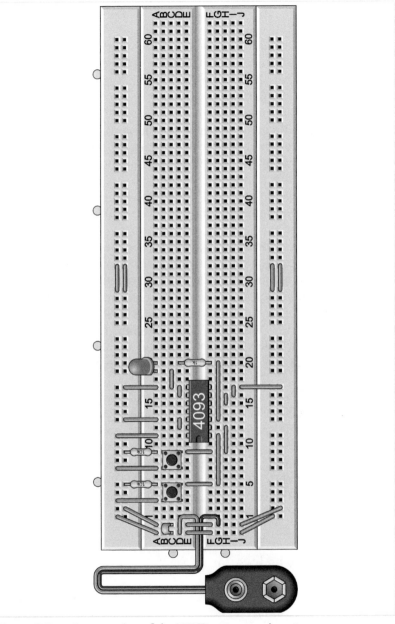

Figure 3.7   Construction of the NAND gate experiment

25

Input 1	Input 2	Output
0	0	1
0	1	1
1	0	1
1	1	0

Table 3.1   The NAND function

Test all four possible input states by operating the two pushbuttons and compare the output states with those in the table. The result is clear: The LED is off only when both switches are pressed. We can also create the NAND function by combining AND and NOT. On the AND output, the state would only be '1' when both inputs are in the '1' state. The inverter following the AND gate reverses the output state.

If the result of the experiment is different to what we expected, it's time for some troubleshooting. A voltmeter is especially suited for this, and an oscilloscope is even better for experiments that are more complex. You should measure all voltages on the individual IC pins against GND (ground). Before anything else, check the ground pin, 7 (GND, 0 V), and pin 14 (VCC, +9 V). Then examine the relevant inputs. When you operate the connected switches, the voltages should change. If not, the fault probably lies in the connections.

You can also test the output with the voltmeter. If you measure a '1' level (9 V), but the LED still doesn't light, the fault could be an incorrectly connected LED. The cathode is the short lead, and the package also has a flat edge on the cathode side, which is clearly visible in drawings.

Instead of a voltmeter, you could also use an LED with a 1 kΩ resistor in series for fault-finding. Leave the cathode connected to GND and use a connected length of wire on the other end to scan individual test points.

3.2   NAND gate as inverter

One often needs an inverter (NOT gate) in a circuit, but only has an unused NAND gate available. No problem – a NAND gate works as an inverter. You need only connect one of its inputs permanently to VCC.

NAND GATE AS INVERTER 3.2

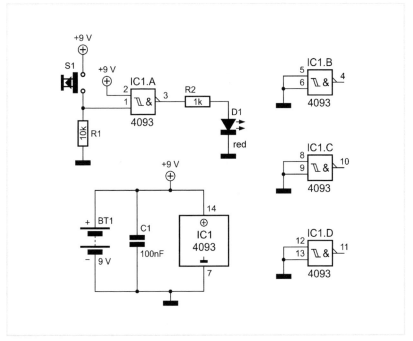

Figure 3.8 NAND gate as inverter

The other input is then the input of the NOT gate. To test, you need only connect a switch.

The truth table shows how a NAND gate can become an inverter. Because Input 2 has a constant '1' level, there are only two different states, depending on Input 1. The output is the inverse of the input. The original NAND truth table had four rows, in accordance with the four possible values of a two-bit binary input number. Input 2 originally had the states '0', '1', '0', '1', but is now held constantly in the '1' state. Therefore, there are only two different input states, corresponding with those of an inverter.

Input 1	Input 2	Output
0	1	1
1	1	0

Table 3.2 NAND gate as inverter

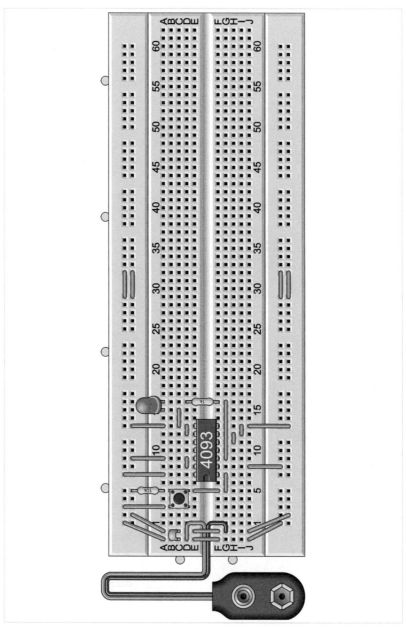

Figure 3.9 Construction of the inverter

NAND GATE AS INVERTER 3.2

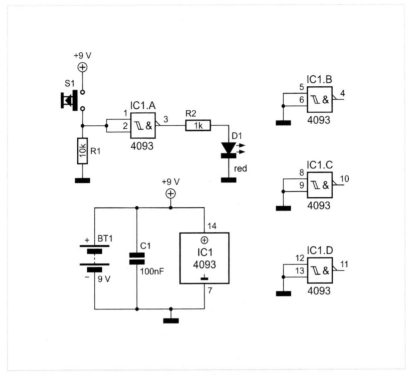

Figure 3.10 NAND gates with adjacent inputs connected together

There are still other ways to use a NAND gate as an inverter. You could combine both inputs into one by connecting them together. This is a popular method, and widely used, because connecting two adjacent pins together is always easily achieved on a circuit board.

The table of input states looks a little different, but the result is the same: Still an inverter. We can clearly see that, from the four rows of the NAND truth table, still not more than two different possible states remain.

Input 1	Input 2	Output
0	0	1
1	1	0

Table 3.3 NAND gate with connected inputs as inverter

29

3 GATE FUNCTIONS

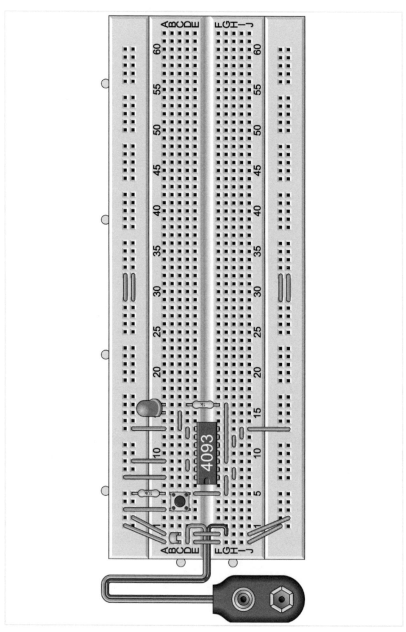

Figure 3.11 *Construction of the inverter (2)*

3.3 AND gate

Logically, the NAND gate consists of an AND gate and an inverter. One may reverse the result with an additional inverter, and return to the AND function. As demonstrated in the previous section, the inverter is realised in the form of yet another NAND gate. So, an AND gate is made from two NAND gates.

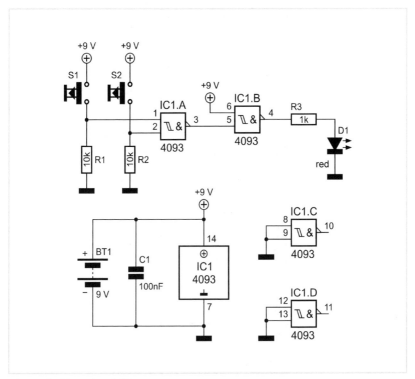

Figure 3.12 An AND gate

Input 1	Input 2	Output
0	0	0
0	1	0
1	0	0
1	1	1

Table 3.4 The AND function

3 GATE FUNCTIONS

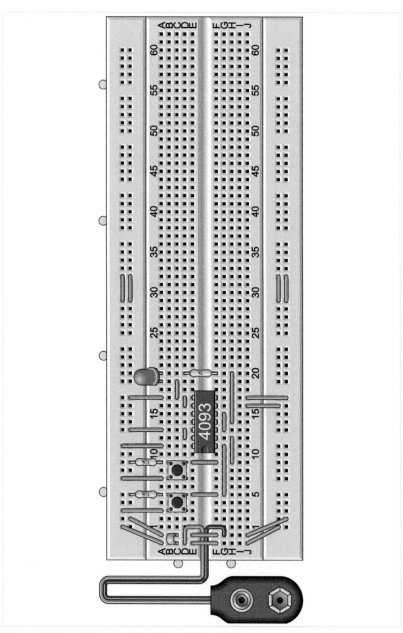

Figure 3.13   *Construction of an AND gate*

3.4 NOR gate

Now, we introduce another IC: The 4001 has four NOR gates, i.e. OR gates with inverters at their outputs. This experiment demonstrates the behaviour of a single NOR gate:

The result: The LED lights only when neither of the two switches is depressed.

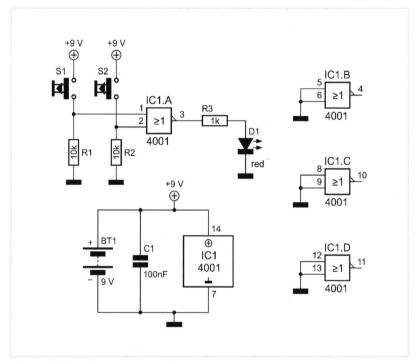

Figure 3.14   NOR gate experiment

Input 1	Input 2	Output
0	0	1
0	1	0
1	0	0
1	1	0

Table 3.5   The NOR function

3 GATE FUNCTIONS

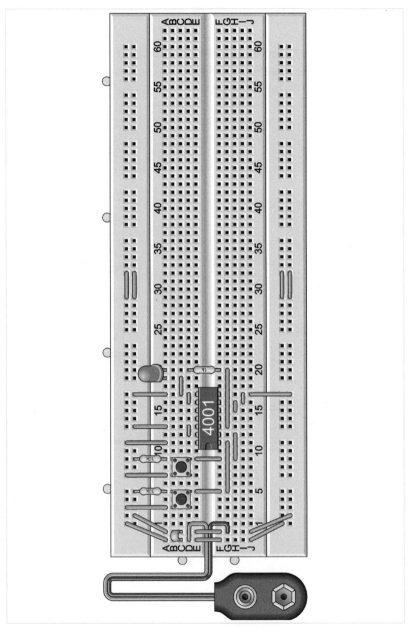

Figure 3.15 Construction of the NOR gate

3.5 NOR gate as inverter

An inverter can be built from a NOR gate in exactly the same manner as from a NAND gate. There are again two ways to attack this problem. The first way is to tie one of the two pins to a fixed level, but note: This time, it must be the '0' level.

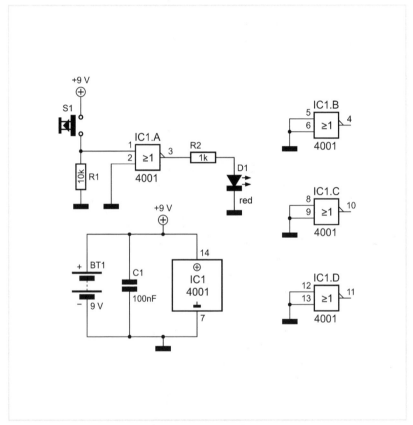

Figure 3.16 An inverter from a NOR gate

Input 1	Input 2	Output
0	0	1
1	0	0

Table 3.6 NOR gate as inverter

3 GATE FUNCTIONS

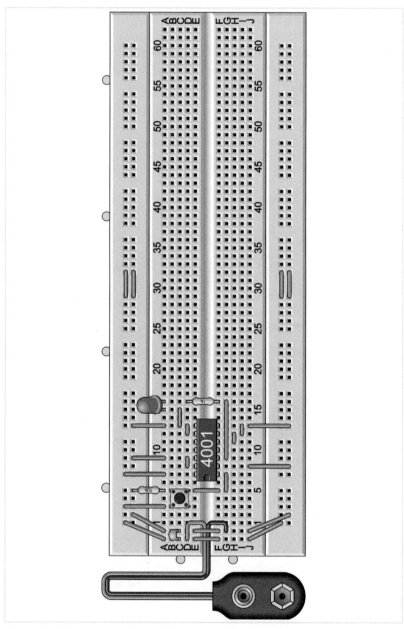

Figure 3.17   Construction of the inverter

3.5 NOR GATE AS INVERTER

The second variant of the circuit using NOR gates looks exactly like that using the NAND gate. Connect both inputs of the NOR gate together, and the inverter is complete.

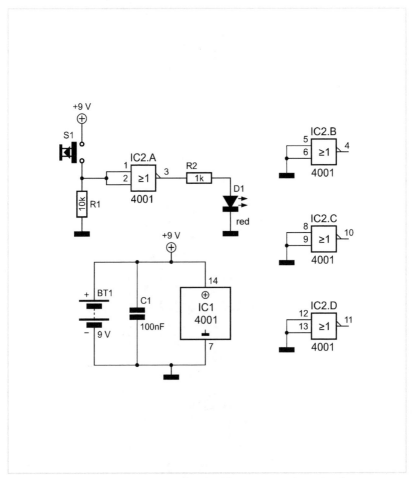

Figure 3.18   *An inverter made from a NOR gate with bridged inputs*

Input 1	Input 2	Output
0	0	1
1	1	0

Table 3.7   *NOR gate with bridged inputs as inverter*

3 GATE FUNCTIONS

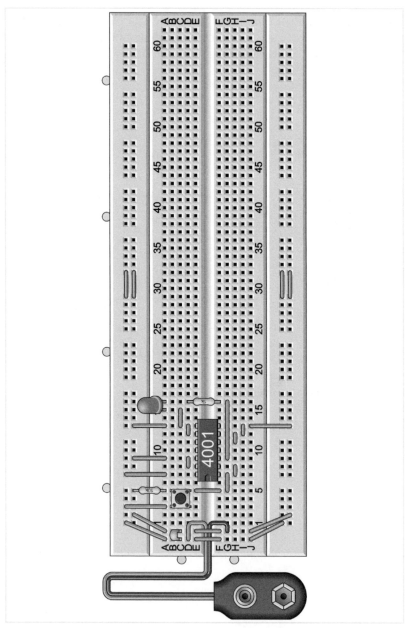

Figure 3.19   *Construction of the inverter with bridged NOR inputs*

3.6 OR gate

Logically, a NOR gate consists of the OR and NOT basic functions. Once again, the internal inverter must be cancelled out to create an OR gate. Connect an inverter after the NOR gate to create the OR gate.

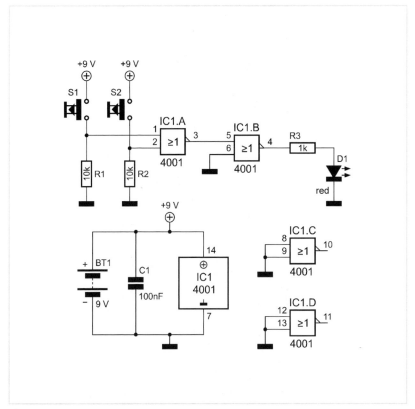

Figure 3.20 An OR gate

Input 1	Input 2	Output
0	0	0
0	1	1
1	0	1
1	1	1

Table 3.8 The OR function

3 GATE FUNCTIONS

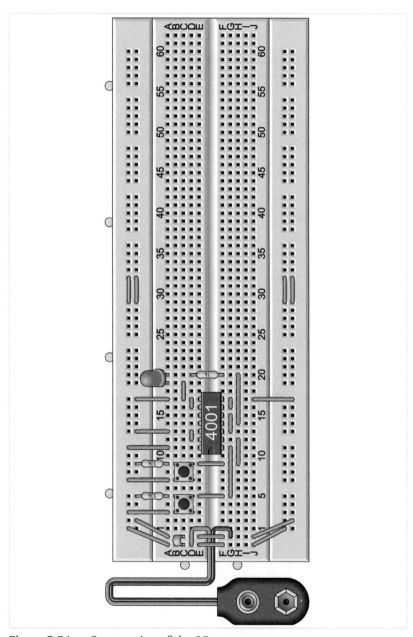

Figure 3.21   *Construction of the OR gate*

3.7 NOR becomes AND

Now it gets more complicated. Let's construct an AND gate from a few NOR gates. The solution consists of first inverting the two input states before connecting them to the input of the NOR gate.

By comparing the truth table of an AND gate with that of an OR gate, it is clear why we end up with the AND function. When you invert the in-

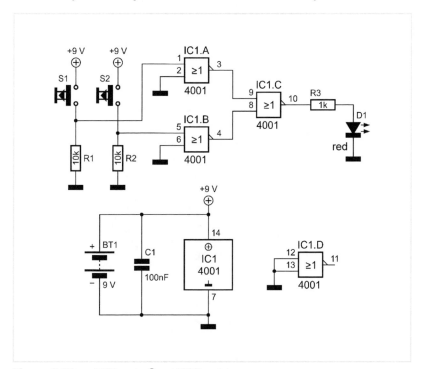

Figure 3.22   AND gate from NOR gates

Input 1	Input 2	Output
0	0	1
0	1	0
1	0	0
1	1	0

Table 3.9   The NOR function

Input 1	Input 2	Output
0	0	0
0	1	0
1	0	0
1	1	1

Table 3.10   The AND function

puts of the NOR gate, the AND function's result is output - note that the order of the input states changes when we invert them. Note the top row in the NOR table (inputs '00') and the bottom row of the AND table (inputs '11').

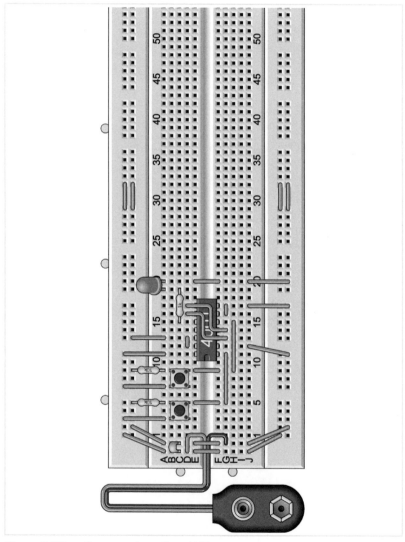

Figure 3.23   Construction of an AND gate

3.8 NOR becomes NAND

The AND function may be implemented using three NOR gates, leaving one free. Use it as an inverter and place it at the output of the self-built AND gate. The result is a NAND gate.

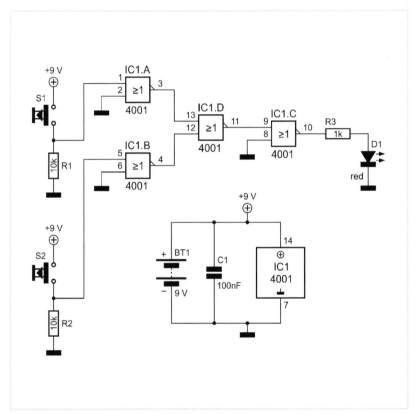

Figure 3.24   A NAND gate

Input 1	Input 2	Output
0	0	1
0	1	1
1	0	1
1	1	0

Table 3.11   The NAND function

3 GATE FUNCTIONS

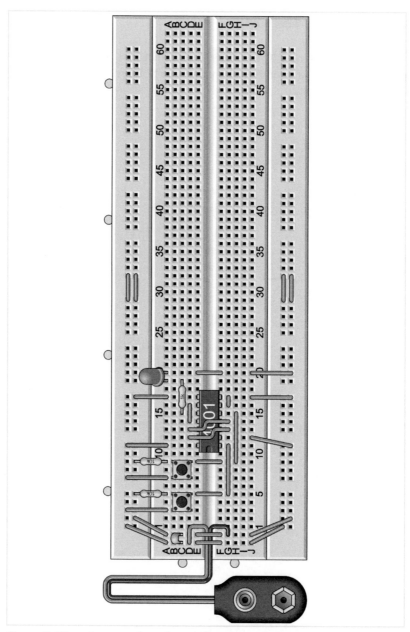

Figure 3.25   *Construction of a NAND gate*

3.9 NAND becomes OR

And now, the same thing in reverse: If we can build an AND gate from NOR gates, we can create an OR gate from NAND gates. Again, the inputs are inverted.

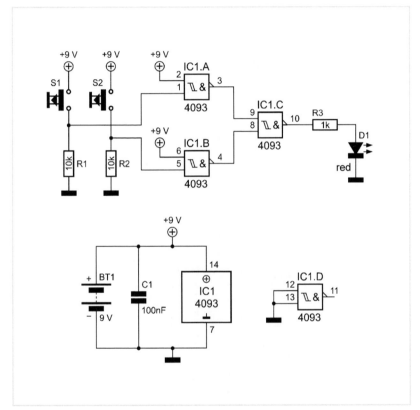

Figure 3.26   The OR gate

Input 1	Input 2	Output
0	0	1
0	1	1
1	0	1
1	1	0

Table 3.12   The NAND function

Input 1	Input 2	Output
0	0	0
0	1	1
1	0	1
1	1	1

Table 3.13   The OR function

3 GATE FUNCTIONS

Comparison of the two functions shows what's happening here. With a NAND gate, the output is only '0' when both inputs are '1'; in contrast, the OR gate outputs '0' only when both inputs are '0.' A NAND gate may thereby be transformed into an OR gate when both inputs are inverted.

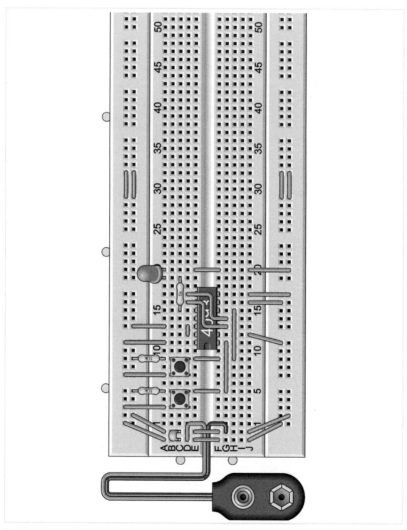

Figure 3.27   *Construction of an OR gate*

3.10 NAND becomes NOR

If you take the previous circuit and invert the result once again, the self-made OR gate becomes a NOR gate. Compare the circuit with that in Section 3.8. The similarity should not surprise you.

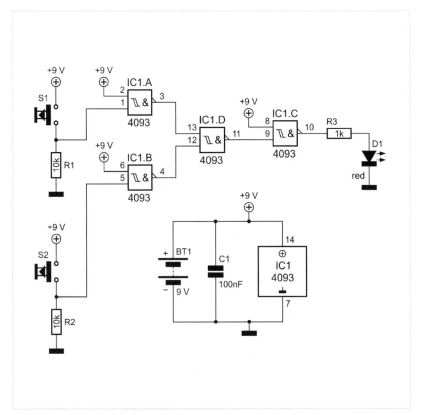

Figure 3.28   NOR gate from NAND gates

Input 1	Input 2	Output
0	0	1
0	1	0
1	0	0
1	1	0

Table 3.14   The NOR function

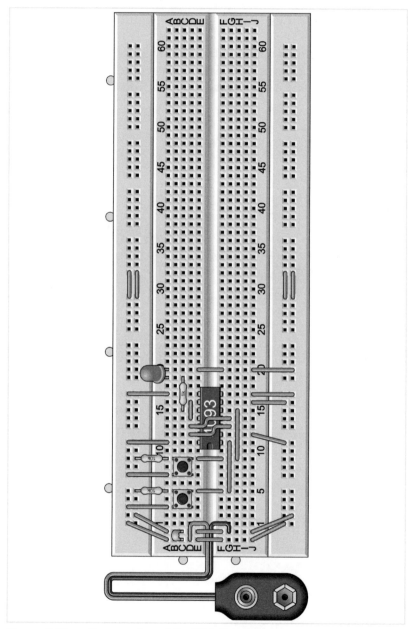

Figure 3.29   *Construction of the NOR function*

3.11 XOR

The antivalence function is also known as exclusive-OR (XOR). Its output is '1' when its inputs are in different states. When both are '1' or both are '0', the output is 0.

Compare your results with the XOR function's truth table. The experiment shows than an XOR function actually has been created, even though this may not be so clear from the circuit. To understand it pro-

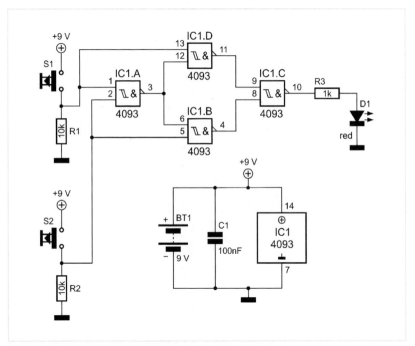

Figure 3.30   An XOR gate

Input 1	Input 2	Output
0	0	0
0	1	1
1	0	1
1	1	0

Table 3.15   The XOR function

3 GATE FUNCTIONS

perly, try the following: Sketch the circuit and note any input combination, e.g. '01'. The gate's output state can then be determined. The output state of Gate A ('1') is connected to the inputs of Gate D ('1', '1') and the output of D is '0', and so on. When you think through the circuit from left to right, the result is as desired.

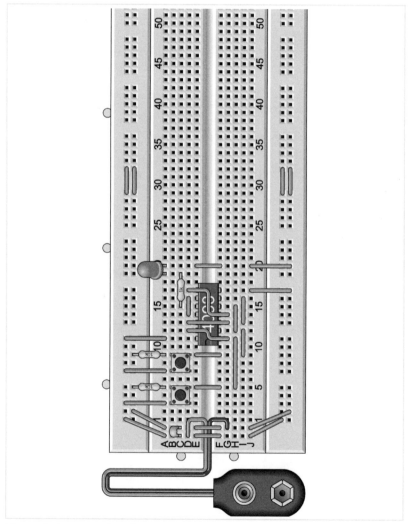

Figure 3.31   *Testing the XOR gate*

3.12 XNOR

The equivalence function (XNOR) is the inverse of the XOR function. Its output is '1' whenever the inputs are the same (equivalent). All that's required is to invert the output of the XOR circuit. However, in the previous circuit, we used all of the gates in the 4093, so one IC will no longer suffice. While there are several possible solutions to this, an

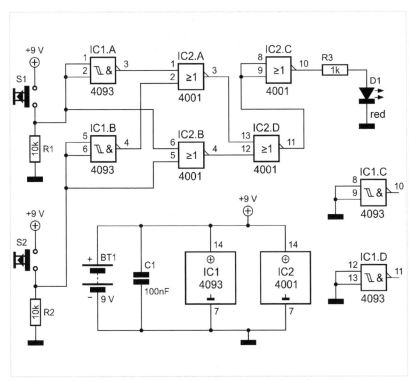

Figure 3.32   An XNOR gate

Input 1	Input 2	Output
0	0	1
0	1	0
1	0	0
1	1	1

Table 3.16   The XNOR function

3 GATE FUNCTIONS

unusual approach is taken here – the XOR gate is created from a NOR gate, using two NAND gates as inverters. Finally, another inverter follows, to turn the XOR function into an XNOR.

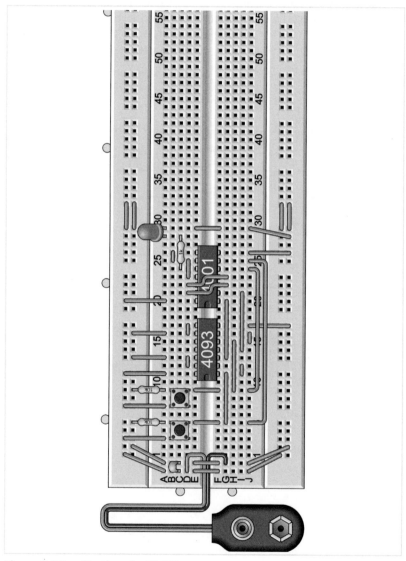

Figure 3.33   *Testing the XNOR gate*

3.13 The majority function

In a democratic election, votes must be counted. When a majority has voted "yes", the result should be "yes". The result is always clear if the number of votes is odd. An even number of participants might result in a tie.

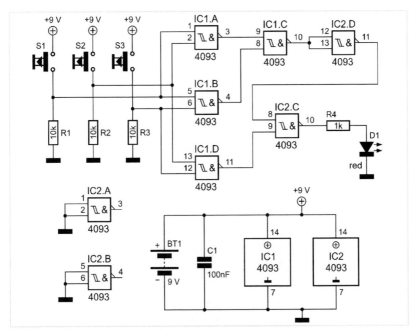

Figure 3.34   *Illustration of the majority function*

Input 1	Input 2	Input 3	Output
0	0	0	0
0	0	1	0
0	1	0	0
0	1	1	1
1	0	0	0
1	0	1	1
1	1	0	1
1	1	1	1

Table 3.17   *The majority function*

3 GATE FUNCTIONS

In the majority function, the output state corresponds with the majority of the input states. Obviously, this will not work with only two inputs, as we need an odd number of inputs. So, we construct a circuit using three inputs. If at least two of the inputs are '1', the output is '1'. On the other hand, if at least two of the inputs are '0', the output is '0'.

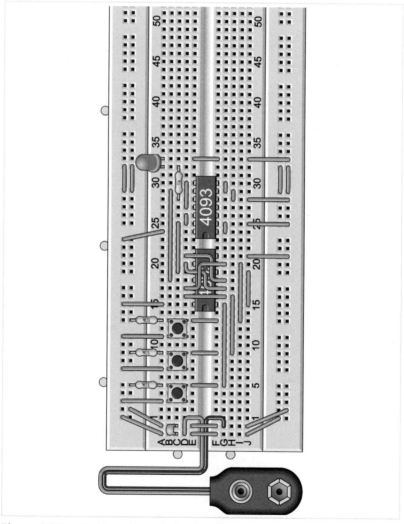

Figure 3.35   *Testing the majority function*

4 Flip-Flops

A flip-flop is a logic circuit whose output state depends not only on the current state of its inputs, but also on its previous state. Under certain conditions, the initial state will be toggled from '0' to '1', or vice-versa.

Flip-flops are also a form of memory. A one-bit memory retains its '0' or '1' state indefinitely, and its state can be changed at will. In the previous chapters we discussed circuits with gates; now we will discuss circuits with memory elements.

4.1 RS flip-flop from NOR gates

An RS flip-flop is a circuit whose output state, '0' or '1', is stored. It has two inputs, S (Set) and R (Reset), via which one may "set" the output to '1' or "reset" it to '0'. It often has an additional output, whose state is the inverse of the first.

Every flip-flop has a feedback from the output to the input. The state that follows depends, therefore, not only on the inputs, but also on the current output state. The following circuit, using two NOR gates, is built symmetrically. Each output of each gate leads to the input of the other. When neither switch is depressed, either D1 or D2 will be in the '1' state. Upon power-up, it is not predictable which LED will be lit.

Should the output of the top gate be '1', the inputs of the bottom gate will be '0' and '1', and its output therefore '0.' In this situation, the top gate will have the input states '0, 0' and its output state will be '1', which is exactly what would be expected. The feedback sees to it that the existing state is retained. With a short press of S1, we can switch the top output to '0', thus forcing the circuit into its second stable state.

4 FLIP-FLOPS

The truth table indicates the output states for all previous output states. Besides the '0' and '1' states, the 'x' state is used when both inputs are '0.' This is the flip-flop's initial, unknown, state. If either of the two inputs, S1 or S2, is set, the output state is toggled. A special situation arises if both switches are pressed simultaneously. Both outputs will then be '0.'

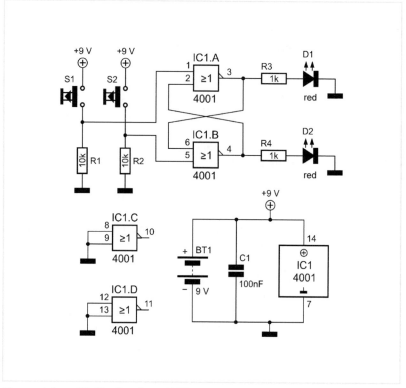

Figure 4.1   *RS flip-flop from NOR gates*

Input 1, S1	Input 2, S2	Output 1, D1	Output 2, D2
0	0	x	NOT x
0	1	0	1
1	0	1	0
1	1	0	0

Table 4.1   *The RS flip-flop*

RS FLIP-FLOP FROM NOR GATES 4.1

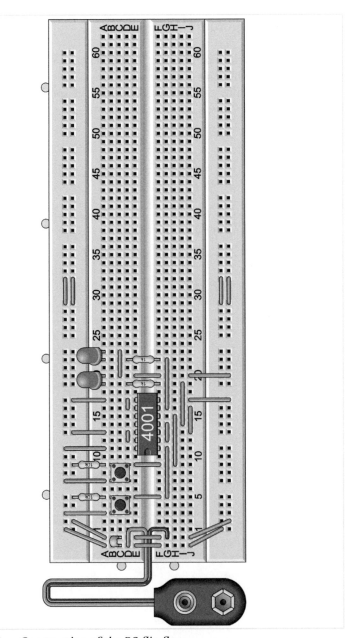

Figure 4.2 *Construction of the RS flip-flop*

4 FLIP-FLOPS

4.2 RS flip-flop from NAND gates

An RS flip-flop can also be constructed from two NAND gates, although it will be in its stable, idle state when a '1' is received at the NAND inputs. If we want a normal '0'-level idle state, the outputs will first have to be inverted. For this reason, both of the 4093's remaining NAND gates are used.

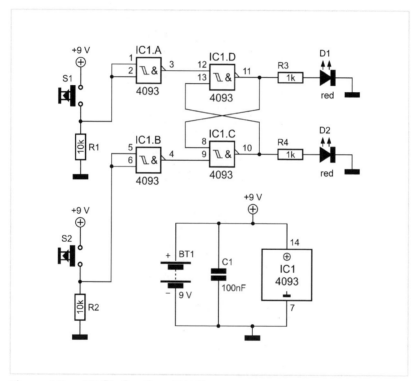

Figure 4.3   *RS flip-flop from NAND gates*

Input 1, S1	Input 2, S2	Output 1, D1	Output 2, D2
0	0	x	NOT x
0	1	0	1
1	0	1	0
1	1	1	1

Table 4.2   *The RS flip-flop with NAND gates*

The truth table corresponds with the original function in the previous section, although, if both inputs are set simultaneously, this time both outputs will be '1'.

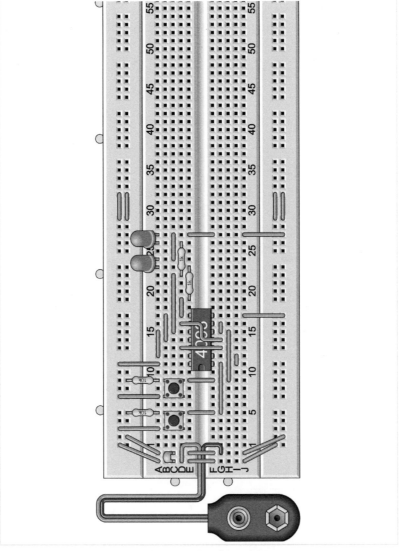

***Figure 4.4**   Construction of the RS flip-flop with NAND gates*

4 FLIP-FLOPS

4.3 JK flip-flop as RS flip-flop

While they can be constructed from individual gates, flip-flops are available as complex devices consisting of many gates. There are many different types, including RS and JK flip-flops. The 4027 contains two edge-triggered JK flip-flops with additional R- and S- inputs. What is achieved with the J and K, as well as C, inputs, is explained in the following section. If these are tied to GND, the circuit behaves just like a

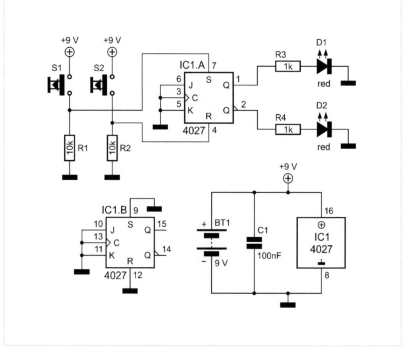

Figure 4.5   *RS flip-flop using the 4027*

Input 1, S1	Input 2, S2	Output Q, D1	Output $\overline{Q}$, D2
0	0	x	NOT x
0	1	0	1
1	0	1	0
1	1	1	1

Table 4.3   *The RS flip-flop in the 4027*

normal RS flip-flop. With a '1' level at S, output Q is set to '1', and reset when there's a '1' level at R. The second output, $\overline{Q}$, is the inverse of the Q. If both inputs are '1', the 4027 behaves like an RS flip-flop made up of NAND gates – just as in the previous section, Q and $\overline{Q}$ are '1'.

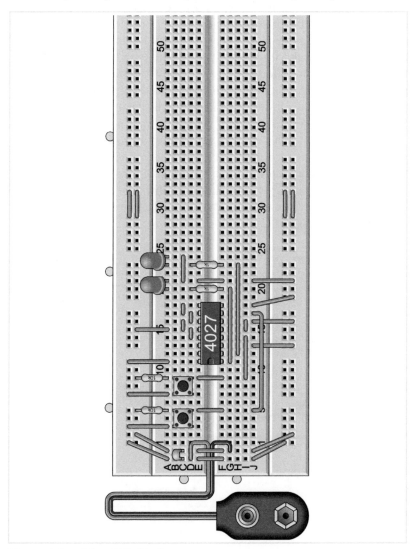

Figure 4.6   *The 4027 flip-flop*

4 FLIP-FLOPS

4.4 The JK flip-flop

A JK flip-flop has three inputs – J, K and C (Clock). To test the basic functionality, the 4027's additional inputs, R and S, must be set to '0'. The outputs, Q and $\overline{Q}$, are set by a positive edge (0→1 transition) at C, depending on the input states, J and K.

An edge-triggered flip-flop is sensitive to the switches' so-called "contact bounce" (also called "chatter"). Upon operation, the contacts touch several times, so that not only one, but several impulses are triggered within approximately a millisecond. Because a keypress should actually produce only one switching edge, a small additional "debounce" circuit must be included to smooth things out. Switch S1 is tied to ground this time, charging capacitor C1 slowly via resistor R2. The 4093's Schmitt trigger produces a clear signal from this. Accordingly, upon release of the button, the capacitor slowly recharges and creates a delayed switch of the output. The keypress then correctly produces only one clock input at C.

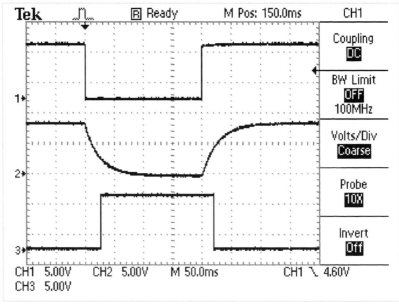

Figure 4.7   *Operation of the clock pulse*

4.4 THE JK FLIP-FLOP

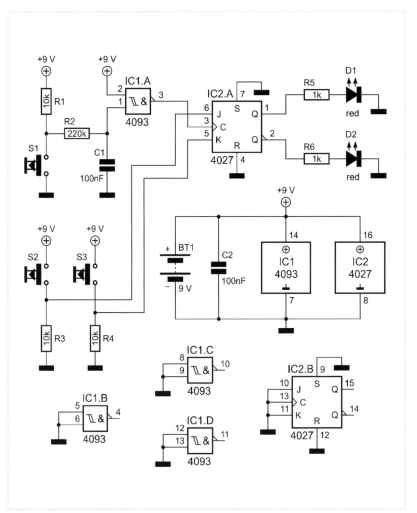

Figure 4.8 Experiment with the JK flip-flop

Input J, S2	Input K, S3	Input C, S1	Output Q	Output Q̄
0	0	transition 0 → 1	x	NOT x
0	1	transition 0 → 1	0	1
1	0	transition 0 → 1	1	0
1	1	transition 0 → 1	toggle	toggle

Table 4.4 The 4027 JK flip-flop

4 FLIP-FLOPS

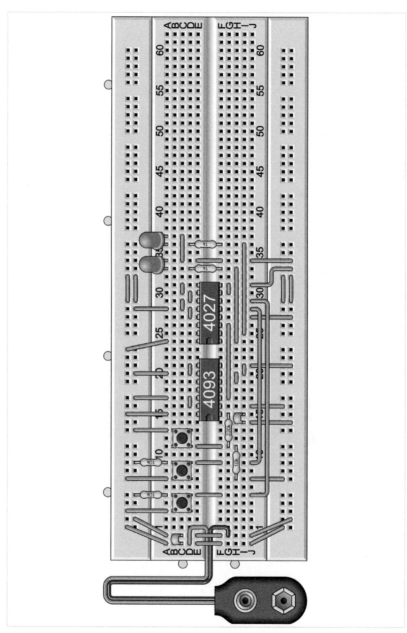

Figure 4.9   4027 JK flip-flop

The truth table illustrates the functionality of the JK flip-flop. The output states each change upon a rising edge at C, depending on J and K. When J and K differ, their states are passed to Q and $\overline{Q}$, respectively, and then remain stored. If J and K stay in the '0' state, the outputs don't change. If both are in the '1' state when the clock edge arrives, the outputs switch from '0' to '1' or '1' to '0' – that is, they are "toggled".

Test all functions of the JK flip-flop. After switching on, either D1 or D2 will be lit, as the initial state of the flip-flop is not predictable. Press only S1, and nothing changes. Now keep S2 depressed and briefly press S1. Now D1 lights up. On the other hand, hold down S3, and LED D2 will light up upon operation of S1. The final test, hold down S2 and S3 simultaneously, and operate S1 a few times. With each press of S1, the outputs are toggled, and thus D1 and D2 are lit alternately.

4.5 Shift registers

A shift register passes the data from stage to stage at each clock pulse. For a four-stage shift register, the results can be seen as follows: At the start, for example, a '1' is at the first output. It is then passed to each stage in four steps.

First, a two-stage shift register is constructed using two JK flip-flops. At each rising edge of the clock signal, a JK flip-flop takes the states at J and K to the output states Q and $\overline{Q}$. It is therefore necessary that J and K are different, which we achieve here using an inverter (IC1.C). The outputs of the first flip-flop (IC2.A) are connected to the inputs of the second flip-flop (IC2.B). Both stages receive the same clock signal. The second JK flip-flop takes on the first flip-flop's prior state. Only at the second clock pulse does the data reach the final output.

Input	Output 1	Output 2	Output 3	Output 4
1	0	0	0	0
0	1	0	0	0
0	0	1	0	0
0	0	0	1	0
0	0	0	0	1

Table 4.5 *The principle of a shift register*

4 FLIP-FLOPS

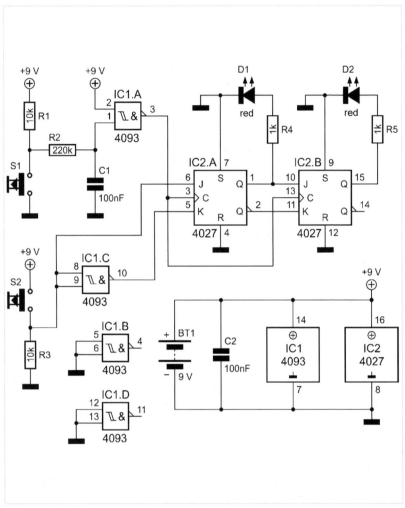

Figure 4.10 *A two-stage shift register*

To test the circuit, hold down S2 and activate the clock pulse using S1. The state of S1 is visible, after a delay, at D1 and then D2. When you keep S2 depressed and press S1 again (Table 4.6), both LEDs light after at most two pulses, because new '1' bits are continuously pushed through. You may also try alternating input states (Table 4.7), so that both LEDs alternate continuously.

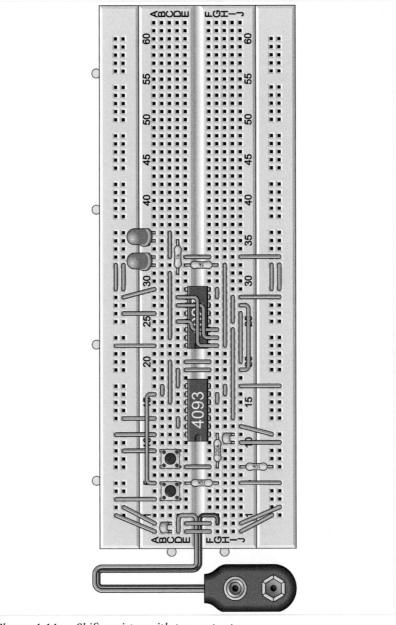

Figure 4.11 Shift register with two outputs

4 FLIP-FLOPS

Input J, S2	Input C, S1	Output Q1	Output Q2
1	transition 0 → 1	1	x
1	transition 0 → 1	1	1
1	transition 0 → 1	1	1
1	transition 0 → 1	1	1

Table 4.6 *Testing the shift register with J = 1*

Input J, S2	Input C, S1	Output Q1	Output Q2
0	transition 0 → 1	0	x
1	transition 0 → 1	1	0
0	transition 0 → 1	0	1
1	transition 0 → 1	1	0

Table 4.7 *Testing the shift register with J = 0/1*

The two-stage shift register is easily expanded using to two additional stages. It then takes four clock pulses for a bit to travel from S2 to D4.

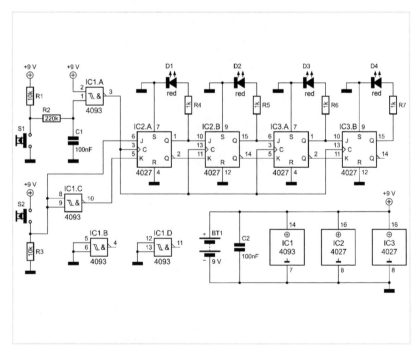

Figure 4.12 *Expanding to four stages*

68

SHIFT REGISTERS 4.5

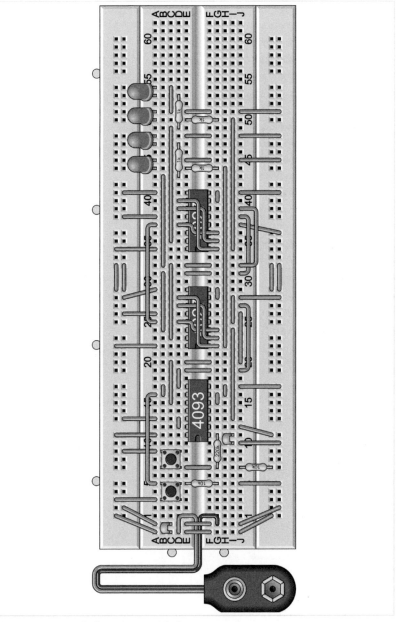

Figure 4.13 Shift register with four outputs

4 FLIP-FLOPS

Input J, S2	Input C, S1	Output Q1	Output Q2	Output Q3	Output Q4
0	transition 0 → 1	0	x	x	x
1	transition 0 → 1	1	0	x	x
0	transition 0 → 1	0	1	0	x
1	transition 0 → 1	1	0	1	0

Table 4.8 *Testing the four-stage shift register with J = 0/1/0/1*

Input J, S2	Input C, S1	Output Q1	Output Q2	Output Q3	Output Q4
1	transition 0 → 1	1	x	x	x
1	transition 0 → 1	1	1	x	x
1	transition 0 → 1	1	1	1	x
1	transition 0 → 1	1	1	1	1

Table 4.9 *Testing the four-stage shift register with J = 1*

4.6 Circular shift registers

In this experiment, four JK shift registers form a ring by feeding the outputs of the final flip-flop to the inputs of the first. Upon power-up, the four stages are found in a random pattern. With every clock pulse (S3), the pattern is shifted by one place. The bit at D4 appears, after a

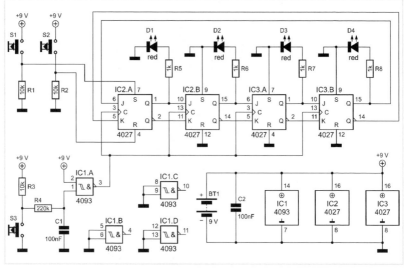

Figure 4.14 *The circular shift register*

CIRCULAR SHIFT REGISTERS 4.6

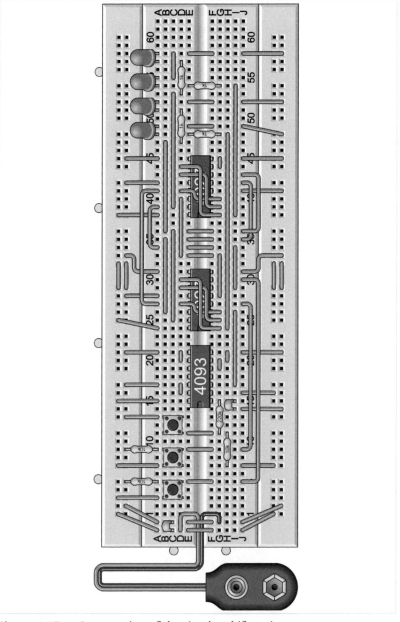

Figure 4.15 *Construction of the circular shift register*

4 FLIP-FLOPS

clock pulse, back at D1. Only the first stage of the shift register may be set (S1) or reset (S2) directly. You may directly input ones or zeros. Imagine, for example, a running light with the state '1,0,0,0'. Once the pattern is entered, only S3 need be used to shift it.

5 Counters

Digital counters are used in clocks, electronic measuring devices and also as elements in the computer field. There are many different types of counters, many of which may be built from 4027 JK flip-flops.

5.1 Counter to 3

Use of a JK flip-flop as a 1-bit counter has already been demonstrated in section 4.3. When both J and K are '1', every positive-going edge of the clock pulse at C toggles the Q output. The output alternates between '0' and '1' states. This can be used for counting. Another way of looking at this is as a frequency divider: Every *two* clock pulses input result in *one* output pulse.

If you feed the output of the flip-flop to the clock input of another JK flip-flop, the number of output pulses is subsequently divided by 4. Because there are four different states, which can be interpreted as a 2-bit binary number, we now have a counter for the numbers from 0 to 3. Because the result is actually a binary number, the inverted output, $\overline{Q}$, must be fed to the next stage's clock input. The next flip-flop toggles upon a positive edge of the clock pulse (0→1 transition), while the transfer of a binary number to the next place takes place at the 1→0 transition that follows.

This circuit is called an asynchronous counter, in contrast to a synchronous counter, in which all clock inputs are connected to the same clock signal and all flip-flops switch at exactly the same moment.

In an asynchronous counter each output controls the next C input. Between input and output, there is always a small time delay of a fraction of a microsecond. The second flip-flop, therefore, switches shortly after the first. For this reason, there exist "invalid" intermediate states.

5 COUNTERS

With a simple counter, this is not a problem, as our eyes would not even be able to distinguish between operations thousands of times slower than that.

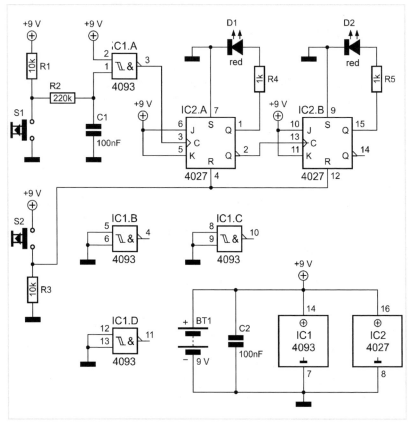

Figure 5.1   *A 2-bit counter*

Input R, S2	Input C, S1	Output Q2, D2	Output Q1, D1
1	0	0	0
0	clock pulse	0	1
0	clock pulse	1	0
0	clock pulse	1	1
0	clock pulse	0	0

Table 5.1   *Operation of the 2-bit counter*

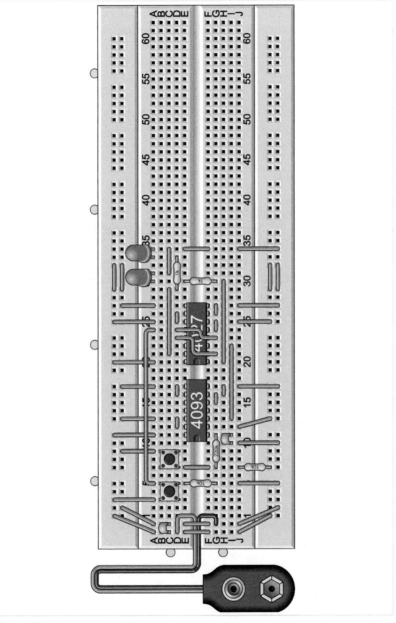

Figure 5.2 Counter with two outputs

5 COUNTERS

The circuit has an additional reset switch (S2). Upon power-up, the circuit is in a random state. Pressing S2 sets the state to '0, 0'. Table 5.1 illustrates the counting sequence at each clock pulse. The circuit counts the clock pulses: 1, 2, 3. The fourth keypress returns the circuit to '0'.

Because the output states are to be read as binary numbers, D1 should be on the right. In the construction of the breadboard, D1 is the LED on the left. If the board is rotated by 180 degrees, however, the binary number is easier to read.

5.2 4-bit counter

The asynchronous 2-bit counter can easily be extended to an asynchronous 4-bit counter with the addition of a second 4027. The circuit then counts from 0 to 15. To read the output states as binary numbers, they must be read from right to left, or just rotate the experiment board. Its function as frequency divider is also easy to see: The individual stages divide the input pulse frequency by 2, by 4, by 8 and by 16, respectively. A pulse appears at D4 after 16 clock pulses.

Input R, S2	Input C, S1	Q4, D4	Q3, D3	Q2, D2	Q1, D1
1	0	0	0	0	0
0	clock pulse	0	0	0	1
0	clock pulse	0	0	1	0
0	clock pulse	0	0	1	1
0	clock pulse	0	1	0	0
0	clock pulse	0	1	0	1
0	clock pulse	0	1	1	0
0	clock pulse	0	1	1	1
0	clock pulse	1	0	0	0
0	clock pulse	1	0	0	1
0	clock pulse	1	0	1	0
0	clock pulse	1	0	1	1
0	clock pulse	1	1	0	0
0	clock pulse	1	1	0	1
0	clock pulse	1	1	1	0
0	clock pulse	1	1	1	1

Table 5.2 Operation of the 4-bit counter

4-BIT COUNTER 5.2

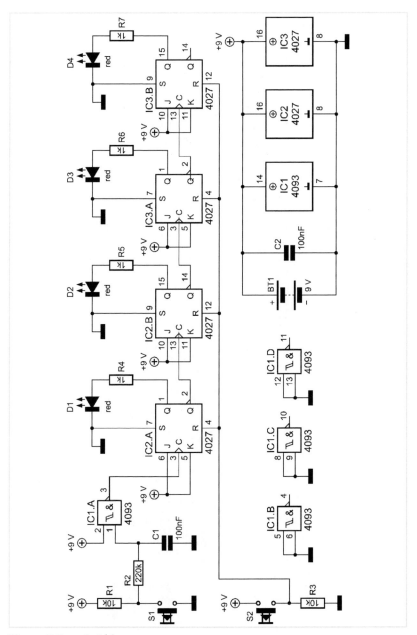

Figure 5.3   *A 4-bit counter*

5 COUNTERS

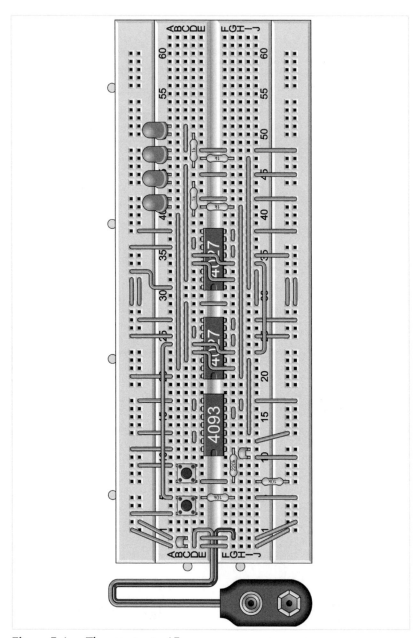

Figure 5.4   *The counter to 15*

5.3 Synchronous counter

The synchronous counter is characterised by the fact that all clock inputs are driven in parallel. All outputs are, therefore, switched at exactly the same time. The result is not visibly different to the asynchronous counter, although gone are the "invalid" intermediate states. This technique is often required when the output states have to control further digital devices. The synchronous counter circuit is significantly more complicated than that of the asynchronous counter. The J and K signals must be prepared so that each flip-flop "knows" ahead of time whether the next clock pulse will require a switch or not. The JK flip-flop toggles when J and K are both '1', and maintains its current state when J and K are both '0.' In the first stage (IC3.A), J and K are permanently set to '1'. The second stage will switch when the Q output of the first stage is '1'. A direct connection between them sees to this.

The third stage will only toggle when both the first and second stages are in the '1' state. Thus, we need to use an AND gate. As already noted

Input R, S2	Input C, S1	Q4, D4	Q3, D3	Q2, D2	Q1, D1
1	0	0	0	0	0
0	clock pulse	0	0	0	1
0	clock pulse	0	0	1	0
0	clock pulse	0	0	1	1
0	clock pulse	0	1	0	0
0	clock pulse	0	1	0	1
0	clock pulse	0	1	1	0
0	clock pulse	0	1	1	1
0	clock pulse	1	0	0	0
0	clock pulse	1	0	0	1
0	clock pulse	1	0	1	0
0	clock pulse	1	0	1	1
0	clock pulse	1	1	0	0
0	clock pulse	1	1	0	1
0	clock pulse	1	1	1	0
0	clock pulse	1	1	1	1

Table 5.3   Operation of the 4-bit synchronous counter

5 COUNTERS

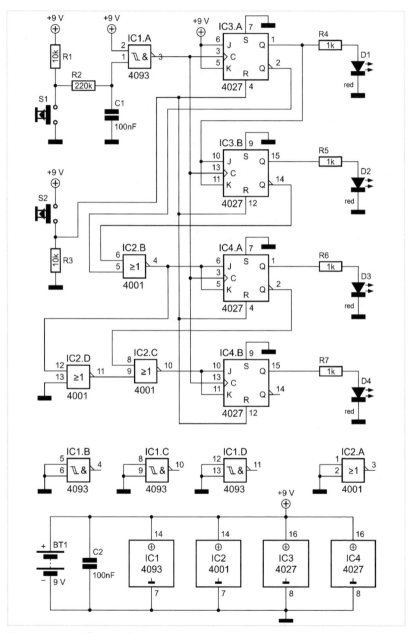

Figure 5.5   The synchronous counter

SYNCHRONOUS COUNTER   5.3

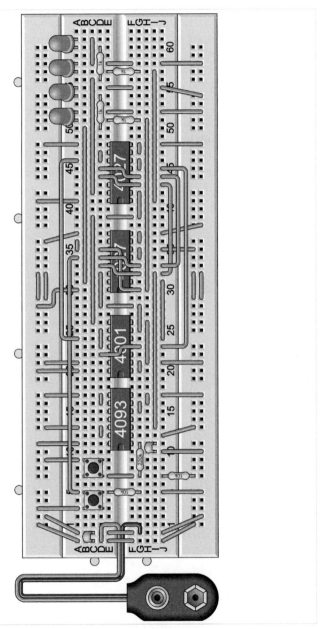

Figure 5.6   *Construction of the synchronous counter*

5 COUNTERS

(Section 3.7), an AND gate can be built from a NOR gate with inverted inputs, so, we've used a NOR gate (IC2.B), to which the inverted outputs of the first two stages are connected. The connection to the $\overline{Q}$ outputs of the first two stages achieves this goal. Accordingly, the two gates, IC2.C and IC2.D connect the $\overline{Q}$ output of the third stage to the output of the first AND gate. As a result, we have an AND gate with three inputs. Only when all three previous stages reach a '1' state, will the fourth stage toggle at the next clock pulse.

5.4 Up/down counter

An up-down counter could be used, as an example, for access control at an event. First, every person entering would pass through a light beam and be counted. At the end of the event, the count direction would be reversed. When all the visitors have left, the counter will be back at zero.

The counter presented here can count up to seven pulses without overflowing, as the three outputs can represent a 3-bit binary number, corresponding to values between 0 and 7. We have again used a synchronous counter, in which all clock inputs are connected in parallel.

Up/down, S2	Input C, S1	Q3, D3	Q2, D2	Q1, D1
0	clock pulse	0	0	1
0	clock pulse	0	1	0
0	clock pulse	0	1	1
0	clock pulse	1	0	0
0	clock pulse	1	0	1
0	clock pulse	1	1	0
0	clock pulse	1	1	1
1	clock pulse	1	1	0
1	clock pulse	1	0	1
1	clock pulse	1	0	0
1	clock pulse	0	1	1
1	clock pulse	0	1	0
1	clock pulse	0	0	1
1	clock pulse	0	0	0

Table 5.4   Operation of the up/down counter

UP/DOWN COUNTER 5.4

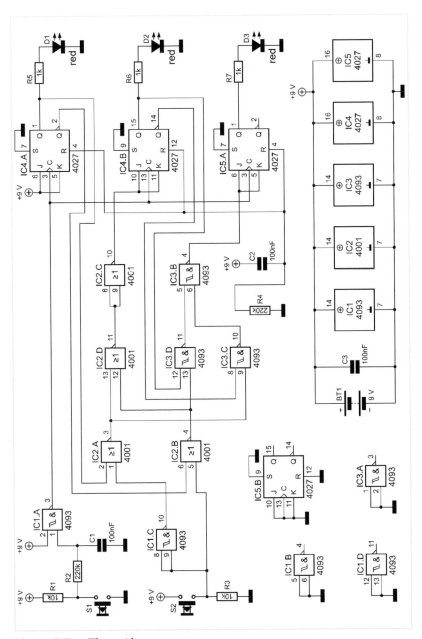

Figure 5.7 The up/down counter

5 COUNTERS

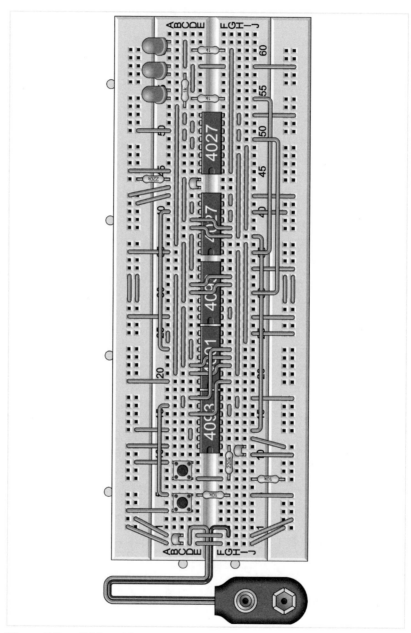

Figure 5.8 3-bit up/down counter

Control of the count direction is done using switch S2. The first counter stage must not be changed, because the '010101' state sets both the upward and downward count. To count downward in the next stage, the counting conditions are reversed. A toggle of the output at the next clock pulse sees to it that J and K reach the '1' state when all previous stages are in the '0' state. A complex circuit consisting of several gates ensures that the state of J and K is inverted when S2 is pressed. This can be verified by testing the levels at IC4.B and IC5.A. Use another LED with a 1-kΩ resistor in series for this.

A search for a reset switch in this circuit will be in vain. Instead, there is an automatic reset circuit consisting of C2 and R4. Upon application of the operating voltage, an approximately 20 ms pulse is triggered. Only when C2 is charged via R4, does a '0' level return to the R input.

6 The Digit Display

While binary '1' and '0' bits are used in digital technology, for everyday use, we use the decimal system, consisting of the Arabic digits, 0 to 9. As a bridge between the two systems, a seven-segment display was developed, which can display all of our conventional digits. Seven LEDs drive seven illuminated bars, from which all of the digits are represented.

6.1 Digit segments

The seven-segment display LD1 contains eight LEDs with a common cathode connection ('cc' at Pin 3 and Pin 8). Seven LEDs are arranged in the display, with optional decimal point. As with normal LEDs, each internal LED requires a resistor in series.

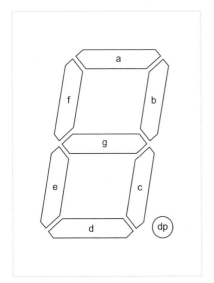

Figure 6.1 Segment layout

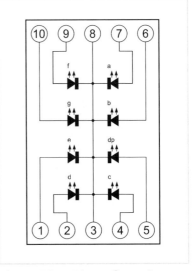

Figure 6.2 Pin configuration

87

6 THE DIGIT DISPLAY

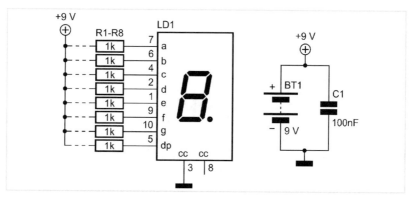

Figure 6.3   Hard-wired seven-segment display

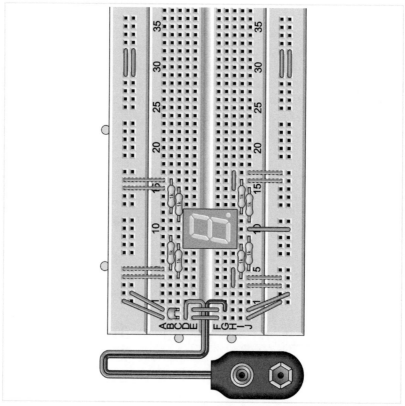

Figure 6.4   Construction of the seven-segment display circuit

The individual segments are indicated by the letters 'a' through 'g.' The uppermost bar is 'a', followed by 'b' to 'f' in clockwise order, and the middle bar, 'g'. First, wire up all of the segments. You will see an '8'. Remove the connection to Pin 10 (Segment 'g'), and a '0' appears. In the same way, you may create all of the digits. The segments 'a', 'b' and 'c' create a '7', 'b' and 'c' a '1', etc. Some letters of the alphabet are even possible. Alternately, you may define special characters.

6.2 Seven-segment decoder

The 4511 seven-segment decoder translates a 4-bit binary number into the bit pattern required to drive a seven-segment display. Construct a 4-bit input using four switches, S1 to S4, and use these to control the four inputs, A to D. Because the outputs are active, LT and BL must be tied to VCC, and EL to GND. The LT (lamp test) input allows for verification that all of the segments actually turn on. Accordingly, BL (blank) turns all the segments off. When EL (enable latch) is in the low state, the 4511 transfers the data on the inputs to its input register.

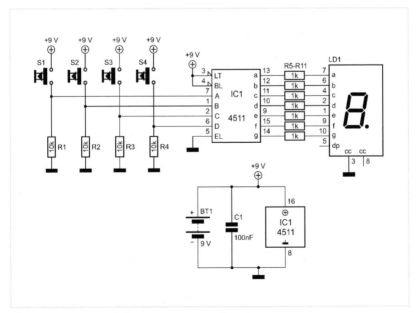

Figure 6.5   *Use of the seven-segment decoder*

Create all possible bit patterns between '0000' (0) and '1001' (9). Values higher than 9 are not displayed. When entering these binary numbers, note that the least-significant bit's switch, S1, is on the left, but represents the right-most binary digit.

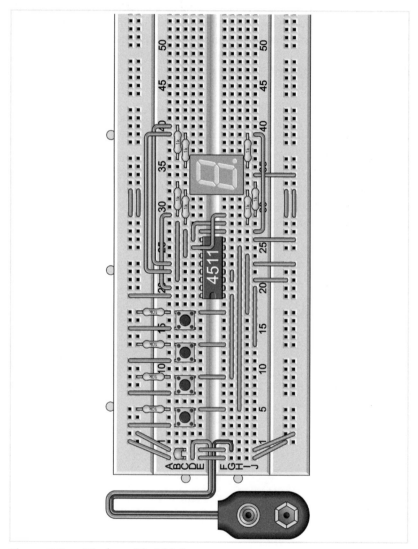

Figure 6.6   *Display with 4-bit input*

6.3 0 to 9 counter

Combine the asynchronous binary counter already presented in section 5.2 with the seven-segment decoder and display. A characteristic of the circuit is that only numbers between 0 and 9 are permitted. For this reason, the number 10 ('1010') is decoded using IC1 and used to create a reset pulse. When the counter is in the '9' state, the next clock pulse will reset the counter to zero. A reset circuit is also created using RC circuit R10/C2 to generate an automatic reset pulse upon power-up, so that the counter always begins at zero. Now, with switch S1, you may count sequentially from 0 to 9.

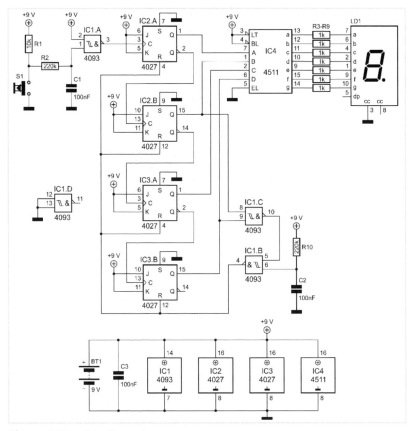

Figure 6.7   *0 to 9 counter*

6 THE DIGIT DISPLAY

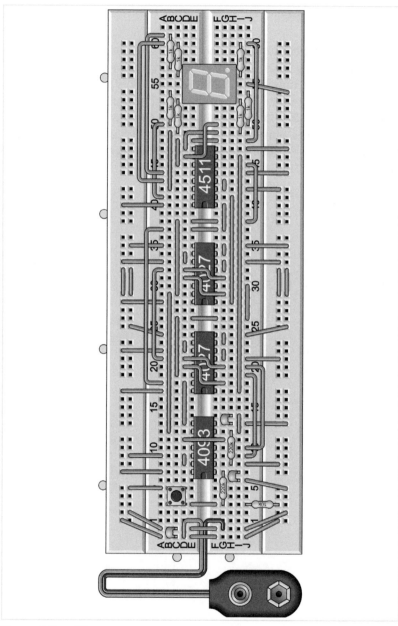

Figure 6.8 The up counter

6.4 9 to 0 countdown

This counter counts backwards from 9 to 0. The 4-bit asynchronous counter is wired so that the Q output, instead of the $\overline{Q}$ output used in the up counter, feeds the next clock input. The reset switch sets the counter to an initial value of '1001' (decimal 9). When a negative overflow of '0000' to '1111' occurs, the circuit is reset to 9. R11 and C3 create a low-pass filter, which prevents a premature reset being caused by the asynchronous counter's short intermediate transition state.

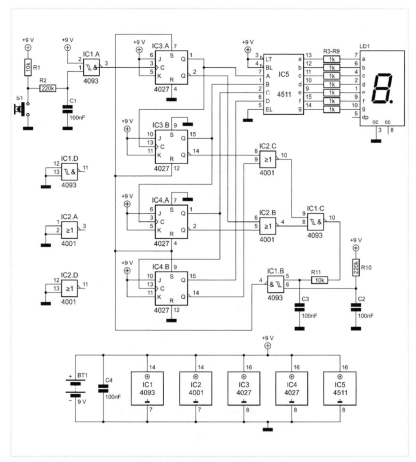

Figure 6.9   *Asynchronous countdown circuit*

6 THE DIGIT DISPLAY

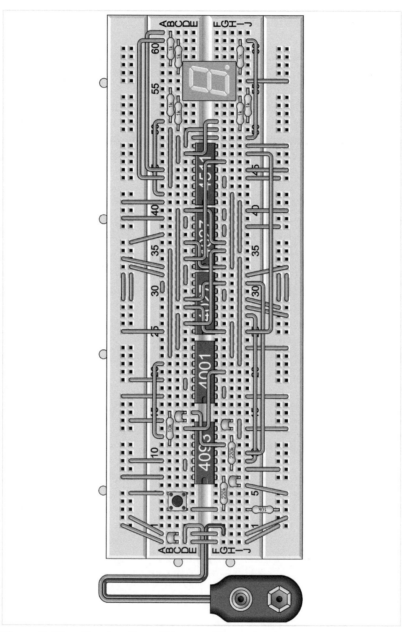

Figure 6.10 *Construction of the countdown circuit*

7 Oscillators

Oscillators are circuits that generate pulsations automatically. A digital oscillator creates a rectangular output signal, which alternates between the '0' and '1' states. Depending on frequency, oscillators may be used to create blinking lights or tone generators.

7.1 Blinking light

The Schmitt trigger in the 4093 has already been used for switch debouncing, but is also suitable for a simple oscillator circuit. The gate's output switches to the '1' state (remember that the output is inverted) when the input voltage dips below 1/3 of the operating voltage (VCC), and returns to '0' only when the input voltage rises above 2/3 of VCC.

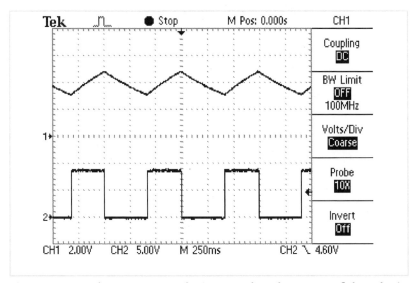

Figure 7.1 *Voltage course at the input and at the output of the Schmitt trigger*

7 OSCILLATORS

With a capacitor on the input and the feedback resistor between the output and the input, a square wave appears at the output. With a 9 V supply voltage, the capacitor is repeatedly charged to 6 V and then discharged to 3 V.

The oscillator's output frequency is dependent on the charge capacitor C1 and the feedback resistor R2. For a rough estimate, the time constant $T = R \times C$ may be used. At 220 kΩ and 10 μF, we get $T = 2.2$ s. The time constant applies, however, to a charge of between 1 and 1/e. Because the oscillator circuit always charges to only between 1/3 and 2/3 of the total voltage, the result is twice as fast. For approximately one second, the oscillator produces a '1' state, and, for the next second, a '0'. From this, we achieve around 30 pulses per minute. The pulse produced by IC1.A is inverted by IC1.B and fed to an LED. The result is a slowly blinking LED.

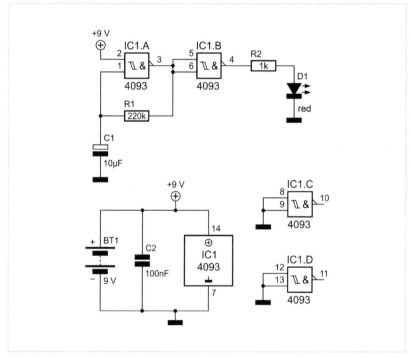

Figure 7.2   Oscillator as blinker

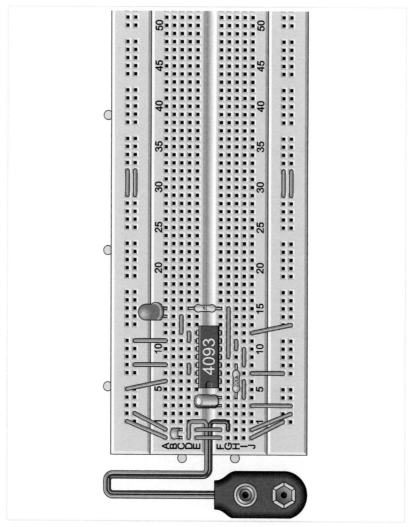

Figure 7.3   Construction of the blinker

7.2   Flip-flop blinker

The blinker circuit is easily expanded to a "flip-flop" blinker consisting of two LEDs. An additional inverter, IC1.C is added. D1 and D2 will now flash alternately.

7 OSCILLATORS

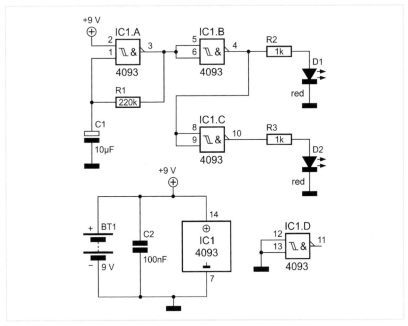

Figure 7.4   *The flip-flop blinker*

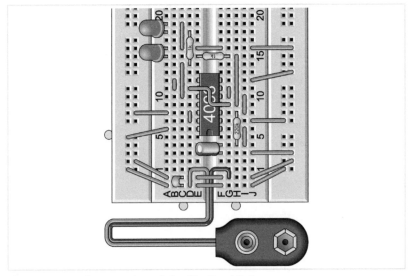

Figure 7.5   *Construction of the flip-flop blinker*

7.3 Metronome

A musician uses a metronome to create a consistent and audible measure of time. For our loudspeaker, we've used a piezoelectric transducer. It may connected directly to a CMOS output without a resistor, because, electrically, it acts as a capacitor with a capacitance of around 20 nF. Current flows only upon a pulse edge, while, with a constant level, the current returns to zero.

This circuit produces approximately 120 level changes per minute. Every change is heard as a click. In order for the sound to be audible, a membrane or resonator is important. Affix the piezo speaker to a piece of cardboard or a wooden surface in order to hear it.

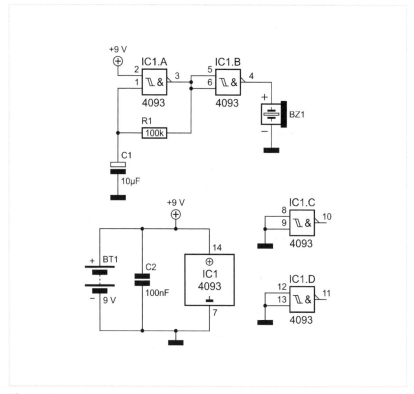

Figure 7.6   *Metronome*

7 OSCILLATORS

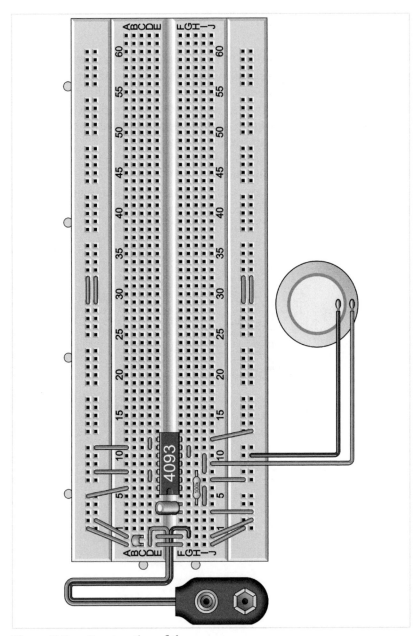

Figure 7.7 Construction of the metronome

7.4 Tone generator

With an RC combination of 3.3 kΩ and 100 nF, the oscillator produces an output frequency of approximately 3 kHz, which is close enough to the piezo element's resonant frequency to be clearly audible. The 4093 NAND gate is used to turn the oscillator on and off. The result: switch S1 serves as a Morse key. Now you have a practical Morse Code practice device.

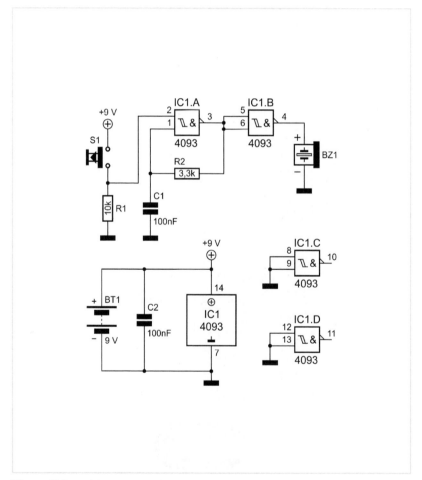

Figure 7.8   *A tone generator*

7 OSCILLATORS

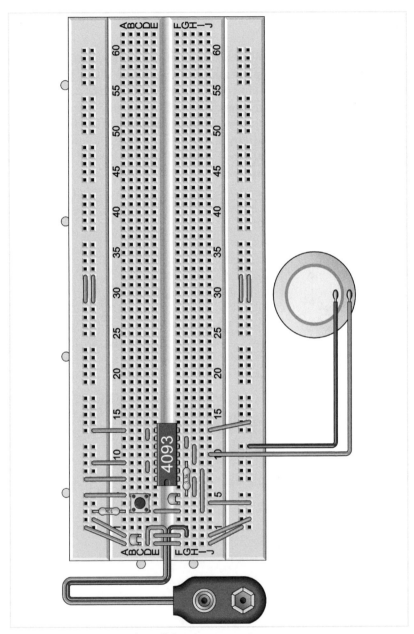

Figure 7.9 Construction of the tone generator

8 Applications

The previous chapters dealt primarily with the fundamentals of digital electronics. Now, practical applications are presented. From familiar basic circuits, countless practical, usable circuits may be developed. We can only show a few examples here. Your own ideas for projects are encouraged.

8.1 Light-controlled tone

With a light-dependent resistor (LDR), the Morse tone generator from section 7.4 is enhanced to become a light-controlled oscillator. The higher the light intensity, the higher the frequency. In full daylight, the

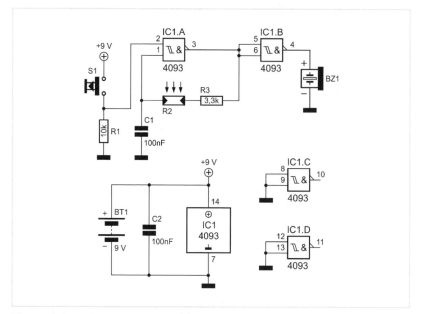

Figure 8.1   *Tone generator with LDR*

103

sensor has a resistance of under 1 kΩ, and, in darkness, several MΩ. The frequency of the oscillator is thus operable through a wide range of brightnesses.

Try play some music by using your hand to cast a shadow. Another interesting application is the investigation of light waves. While natural light changes very slowly, many artificial light sources flicker, which will produce a frequency oscillation in this circuit. One hears a rough, humming tone containing several frequencies. The effect is more noticeable with fluorescent lamps than with filament bulbs.

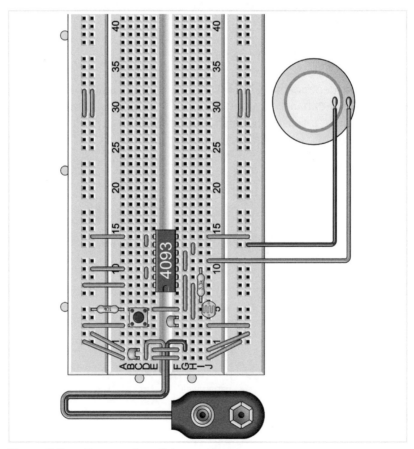

Figure 8.2   Construction of the tone generator

8.2 Mini organ

By changing the feedback resistor's value, you change the circuit's frequency. In principle, we can create a simple electronic organ. The four push buttons, S1 to S4, "switch in" resistances of 10 kΩ, 20 kΩ, 30 kΩ and 40 kΩ. Thus, the organ has four tones, albeit not in any standard musical scale. The circuit merely demonstrates the principle.

Using other resistors, or, even better, potentiometers (variable resistors) and more switches, a very playable organ may actually be built.

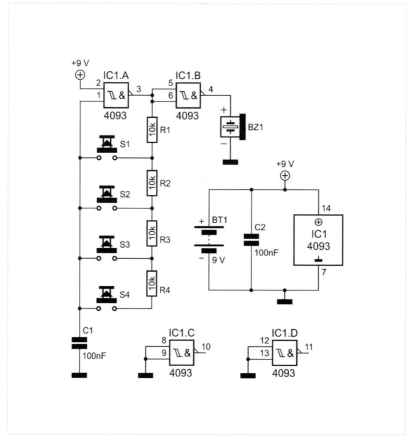

Figure 8.3   *Tone generator with four tones*

8 APPLICATIONS

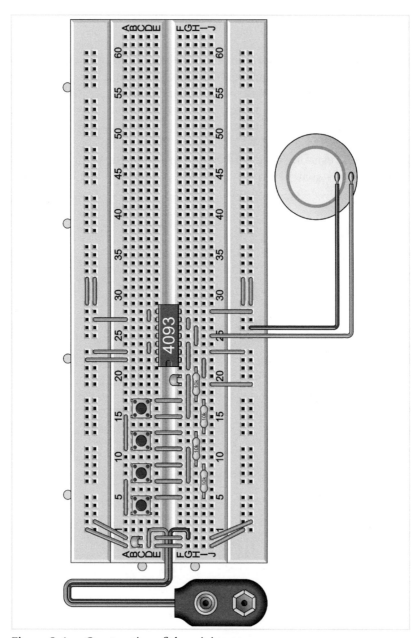

Figure 8.4   Construction of the mini organ

8.3 Siren

The siren produces a tone that alternates between two frequencies. For this, a total of three oscillators is required. IC1.B produces the high tone, and IC1.C the low one. A further oscillator, IC1.A, produces a slow square wave to switch between the two tones.

The circuit consists of four NOR gates. IC2.A acts as an inverter. Either IC2.B or IC2.D will pass a tone on when one of them receives the '0'-level switching signal, thereby activating only one of the two tones at a time. IC2.C combines the two signals.

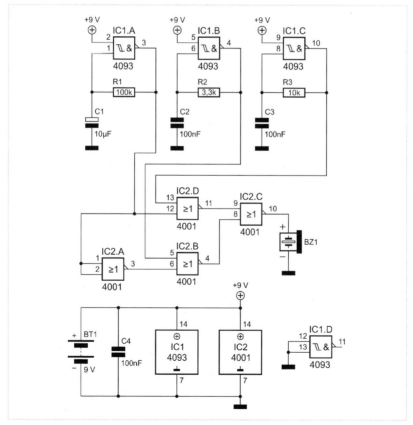

Figure 8.5   *Switching between two tones*

8 APPLICATIONS

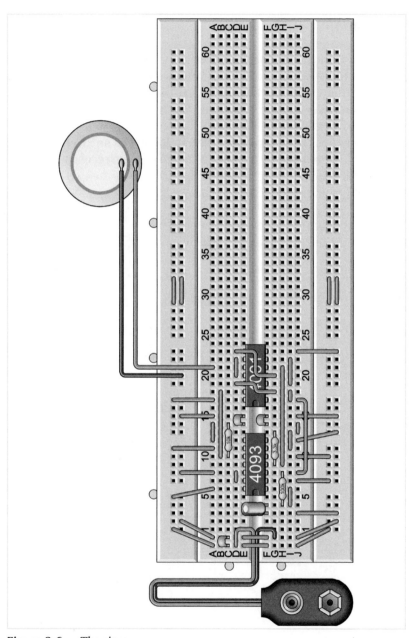

Figure 8.6 The siren

8.4 Twilight switch

The LED in this circuit is turned on when it's dark, and turned off again when it's bright. The result is an automatic night light. At the input of the Schmitt trigger is a voltage divider consisting of the light sensor R1 and a fixed resistor, R2. The voltage reaches all values of between almost zero (darkness) and almost 9 V (bright light). This analogue input signal is used to form an inverted digital signal. The switching hysteresis, that is, the voltage difference between activation and deactivation of the 4093, ensures that the LED does not flicker at intermediate brightnesses.

Should it be desired to modify the switching brightness, R2 may be substituted. Smaller values shift the switching point to higher brightnesses.

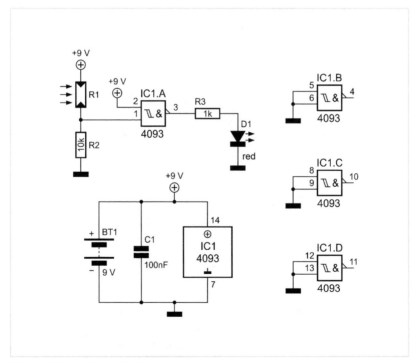

Figure 8.7   The twilight switch

8 APPLICATIONS

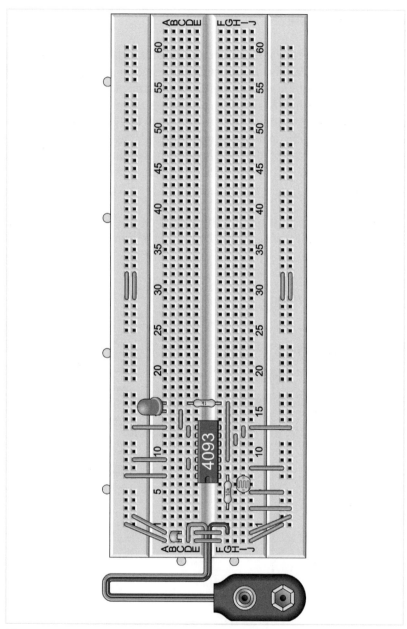

Figure 8.8 *Construction of the twilight switch*

8.5 Alarm system

The siren from section 8.3 may be used in an alarm system. The tone is turned on only when an alarm is triggered. The RS flip-flop, IC1.A, turns on the alarm via NAND gate IC2.B, when Q is in the '1' state. S1 starts the siren, while S2 turns it off. S1 may, instead, be a trip switch or a pressure switch on a door.

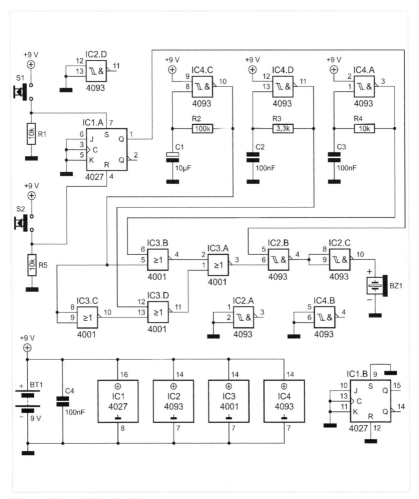

Figure 8.9   Switched siren

8 APPLICATIONS

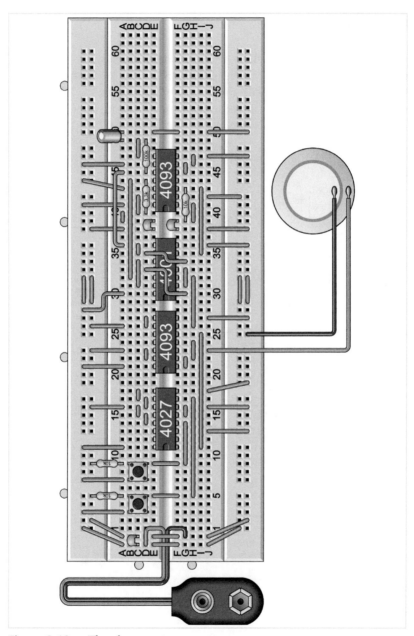

Figure 8.10   *The alarm system*

8.6 Light-activated alarm system

Here, the alarm system in the previous section is modified with the addition of light activation. The LDR, R6, receives illumination when in the idle state. When someone creates a shadow on the light sensor, the voltage at R1 drops and the output of IC2.D switches to '1', causing the alarm to sound. It may be turned off again using pushbutton S1.

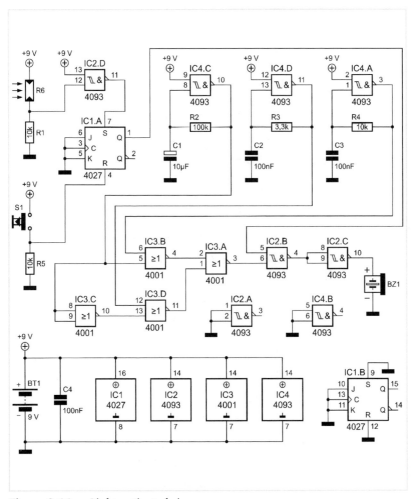

Figure 8.11   *Light-activated siren*

8 APPLICATIONS

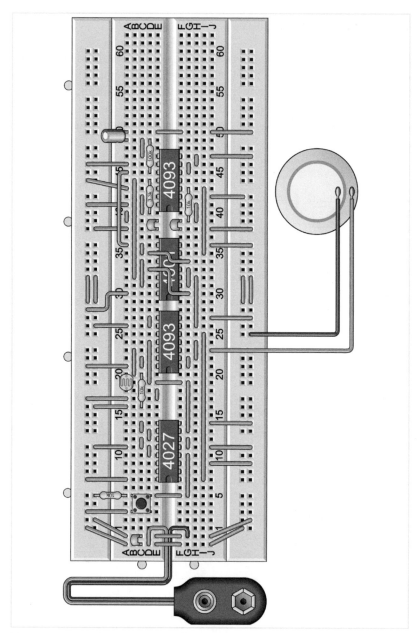

Figure 8.12   *Alarm system with LDR*

8.7 Running light

This running light uses a four-stage shift register, which is controlled by clock pulse generator IC1.A. A special reset circuit produces a defined initial state: IC2.A is set (at S), and all the other stages are reset

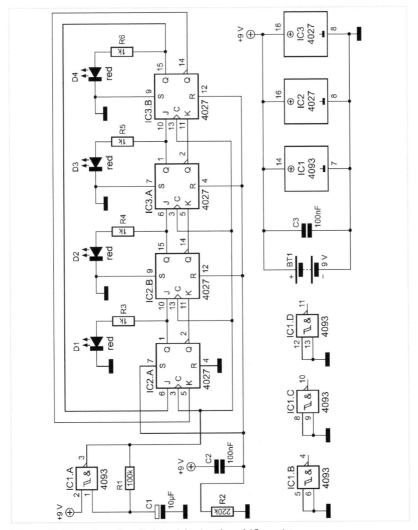

Figure 8.13   *Running light with circular shift register*

8 APPLICATIONS

(at R). Thus, D1 is initially lit. With each clock pulse, the '1' state is passed one stage to the right. Following D4, it's back to D1, as the four stages are wired as a circular shift register.

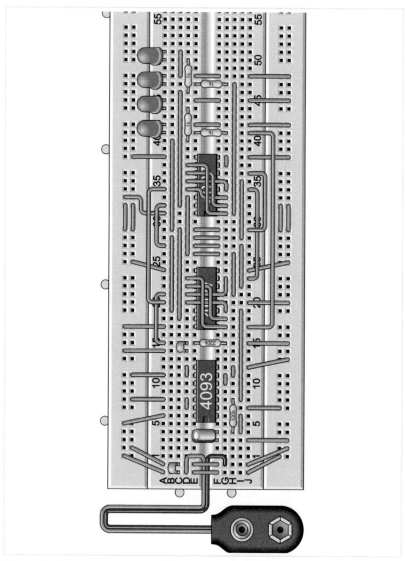

Figure 8.14   Construction of the running light

The running light could be built completely differently, with the same result. This time, an asynchronous 2-bit binary counter is used. The output states run through the binary numbers '00', '01', '10' and '11'. Four NOR gates decode these states. IC3.A switches on only when the input states are '0, 0'. IC3.D is set to the second state '01' – because its upper input is tied to the $\overline{Q}$ output on IC2.A, the NOR gate sees '01' instead of '00'. In the same way, the rest of the decoder gates are connected to the appropriate Q and $\overline{Q}$ outputs of the binary counters.

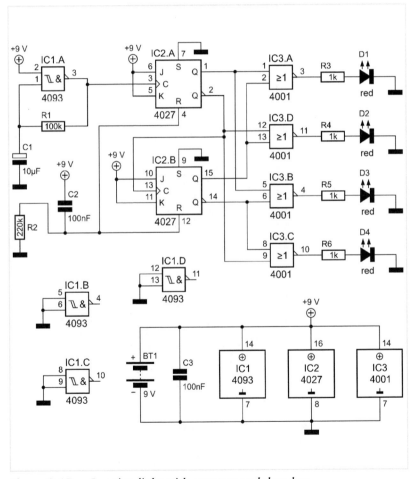

Figure 8.15   *Running light with counters and decoders*

8 APPLICATIONS

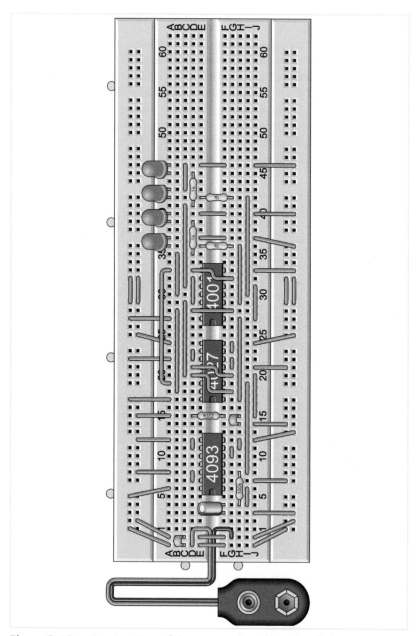

Figure 8.16 Construction of the counter-based running light

8.8 Traffic light control

A traffic light uses three lamps, with the colours green, yellow and red. The control is relatively complex, because different illumination durations are required. The traffic light is red and green for almost equally as long, but there is a short yellow phase before the red phase, and an additional yellow-red phase before the green phase. Our solution is a 4-bit binary counter with a total of 16 states, from '0000' (decimal 0) to '1111' (decimal 15). The decoder that follows switches the connected traffic light LEDs on in the appropriate order. The simplest case is for the red LED. It is on for simply 50% of the time (0 to 7) and off for the other 50% (8 to 15). It may thus be driven directly by the $\overline{Q}$ output of the final stage.

The yellow LED lights up twice during the cycle – in state 7, simultaneously with the red light (briefly before the green phase), and again, alone this time, in state 15 (briefly before the red phase). To decode these two phases, we use an AND gate with three inputs – Bits 0, 1 and 2. It is created using three gates, IC2.C, IC1.C and IC1.B.

Bit 3	Bit 2	Bit 1	Bit 0	Decimal	D1 red	D2 yellow	D3 green
0	0	0	0	0	1	0	0
0	0	0	1	1	1	0	0
0	0	1	0	2	1	0	0
0	0	1	1	3	1	0	0
0	1	0	0	4	1	0	0
0	1	0	1	5	1	0	0
0	1	1	0	6	1	0	0
0	1	1	1	7	1	1	0
1	0	0	0	8	0	0	1
1	0	0	1	9	0	0	1
1	0	1	0	10	0	0	1
1	0	1	1	11	0	0	1
1	1	0	0	12	0	0	1
1	1	0	1	13	0	0	1
1	1	1	0	14	0	0	1
1	1	1	1	15	0	1	0

Table 8.1 *Traffic light truth table*

8 APPLICATIONS

The green LED should light up only when neither the red nor the yellow light is lit. Exactly this logic is taken care of with the NOR gate, IC2.B. Thus, the green LED lights up during states 8 to 14. With a clock frequency of 2 s, the entire traffic light cycle takes 32 s to complete.

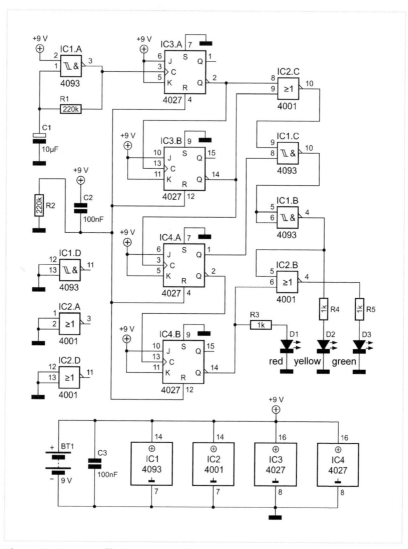

Figure 8.17 Traffic light control

TRAFFIC LIGHT CONTROL 8.8

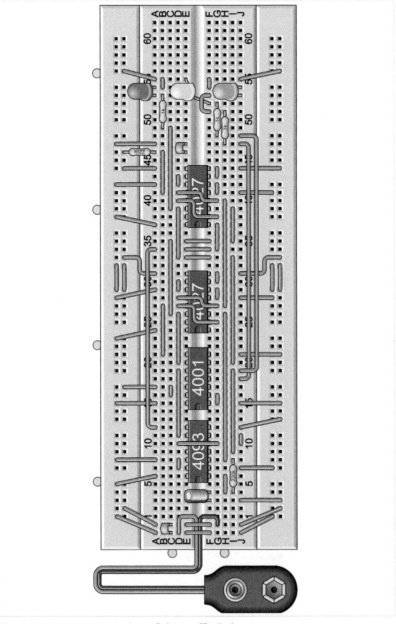

Figure 8.18 *Construction of the traffic light*

8 APPLICATIONS

8.9 Turn-off delay

A turn-off delay is used, for example, in a car's interior lamp. When the vehicle's door is opened, the lamp turns on. Close the door, and the lamp stays on for a few seconds, then turns off. This logic is achieved here using a digital circuit.

A press of S1 charges the capacitor, C1, relatively quickly via D1 and R2. IC1.B turns the LED on. Let go of S1, and current immediately stops flowing through D1, while the capacitor charges very slowly via R3. Only after approximately two seconds, its voltage exceeds the switching voltage of 6 V, so that the LED is turned off.

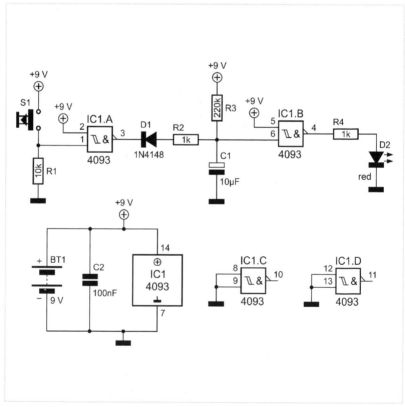

Figure 8.19   *Turn-off delay*

8.9 TURN-OFF DELAY

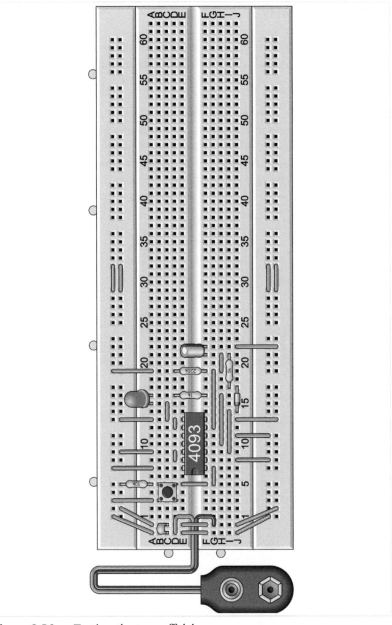

Figure 8.20   *Testing the turn-off delay*

8 APPLICATIONS

8.10 Turn-on delay

With a small modification, the turn-off delay becomes a turn-on delay. D1 is now connected in the opposite direction, R3 leads to ground and C1 to VCC.

In the idle state, S1 is open and the output of IC1.A has a '1' state. D1 conducts, and the input of IC1.B is high, thus turning the LED off. Press S1 and the diode, D1, stops conducting, and the capacitor charges slowly via R3. The input voltage of IC1.B drops, until the LED turns on. Let go of S1, and the LED turns off again almost immediately.

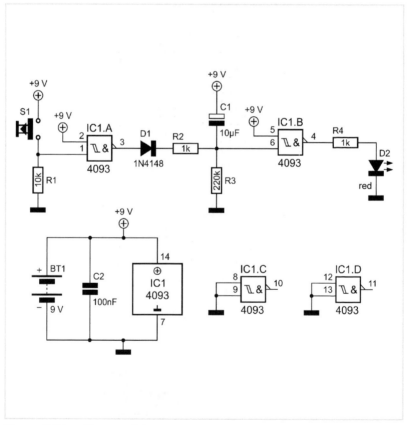

Figure 8.21　　Turn-on delay

TURN-ON DELAY 8.10

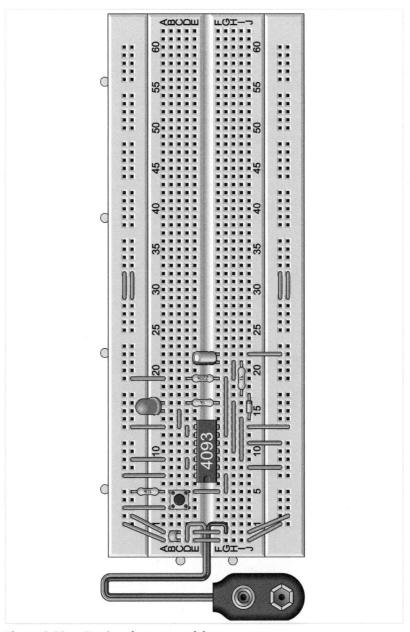

Figure 8.22 Testing the turn-on delay

8 APPLICATIONS

8.11 Time switch

A press of switch S1 produces a pulse of approximately one second, even if the keypress is longer than that. The circuit produces a so-called "monoflop", that is, a single-pulse generator.

A JK flip-flop is used for the purpose. When J = 1 and K = 0, the flip-flop reacts to a positive edge on the clock input, C. The output, Q, then switches to high. At the same time, a delay loop is begun via output $\overline{Q}$. As soon as C2 is discharged, via R5, to approximately 3 V, IC1.C sends a reset pulse to the JK flip-flop. Whether you press the switch briefly or for a long period, the LED, D1, lights up for approximately one second. Press it twice in quick succession, and still only one pulse is produced.

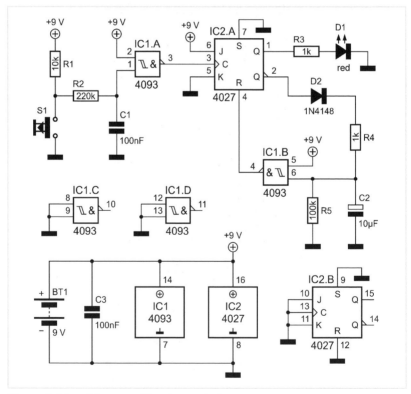

Figure 8.23   *The monoflop time switch*

126

TIME SWITCH 8.11

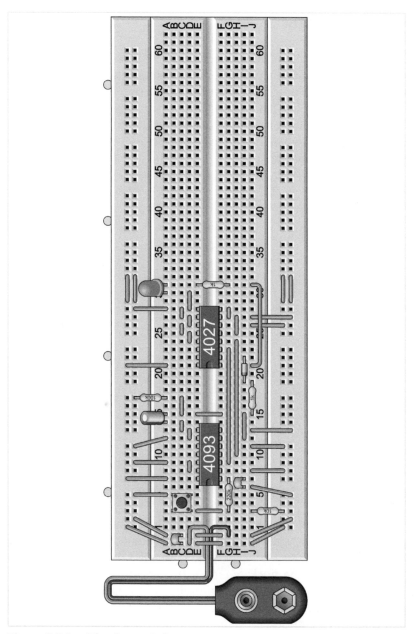

Figure 8.24 *The time switch*

8 APPLICATIONS

8.12 Hallway light timer

A hallway lighting timer should turn the light on for a specified period of time after each keypress. For periods of time longer than about a second, an RC delay loop is not practical, as an especially large capacitor with a very low self-discharge and a very large charge resistor would be necessary. With our available components, a time constant of up to approx. 2 s is possible. To substantially increase the delay, a digital counter circuit is useful.

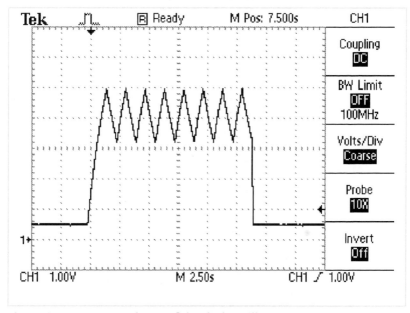

Figure 8.25   Start and stop of the clock oscillator

The circuit presented here uses a clock generator with a period of approx. 2 s. From that, a four-stage asynchronous counter produces a time of approx. 16 seconds. Pressing S1 sets IC3.B to the active state. At the same time, the clock generator is restarted. The first three flip-flops require eight clock pulses in total, until a positive edge appears at the $\overline{Q}$ output of the third stage. This resets IC3.B again. Diode D2 and capacitor C2 are thereby simultaneously discharged. Only at the next start pulse will the clock oscillator begin running again.

HALLWAY LIGHT TIMER 8.12

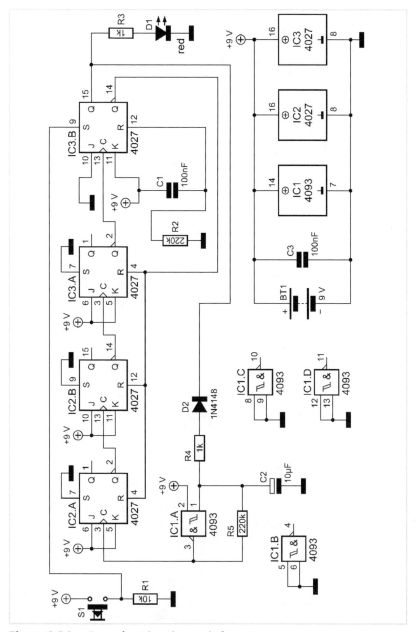

Figure 8.26 Long-duration time switch

8 APPLICATIONS

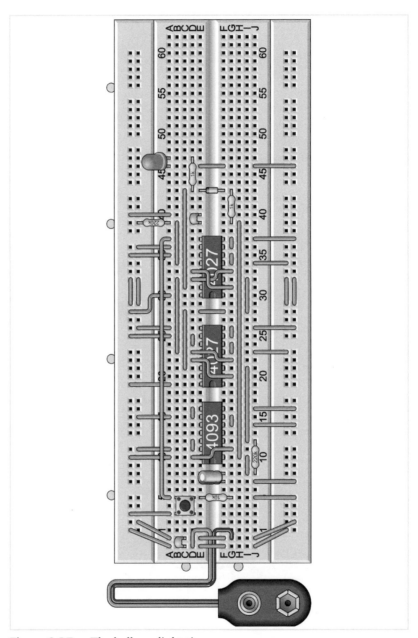

Figure 8.27   *The hallway light timer*

8.13 Simple random generator

Heads or tails, when you flip a coin, both sides appear with equal probability. This random generator serves the same purpose. When you press S1, both LEDs appear to light up simultaneously. As soon as the button is released, only one of the two LEDs remains lit – it is impossible to predict which.

The circuit is based on a clock oscillator at IC1.A and a toggle flip-flop at IC2.A. When S1 is pressed, a clock signal of approximately 10 kHz appears at the clock input of the flip-flop. This frequency is divided by two. An out-of-phase clock signal is produced at Q and $\overline{Q}$ with a frequency of 5 kHz. Because the human eye can't keep up with flashing at

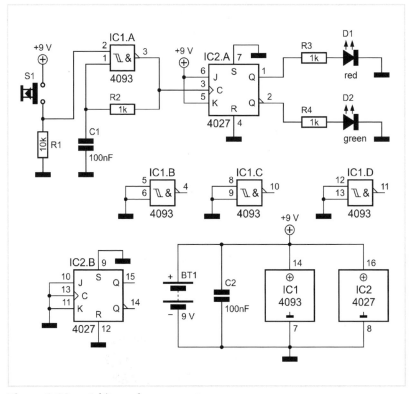

Figure 8.28   *1-bit random generator*

that speed, both LEDs appear lit. As soon as the key is released, the clock signal is halted. The flip-flop is now in one of the two possible states. The result is actually dependent on the exact moment at which the switch is released. Because one cannot have much influence on exactly at which point the switch is released, the result appears random.

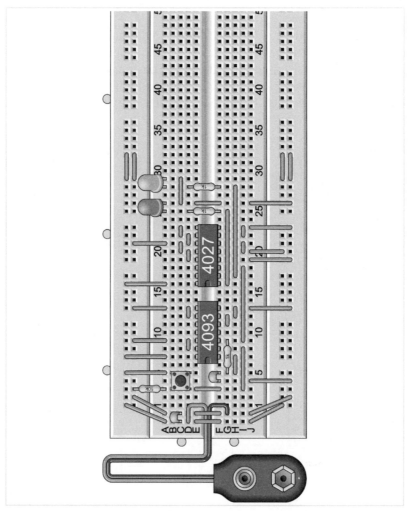

Figure 8.29   *Electronic coin flip*

8.14 Digital roulette

Digital roulette is a further development of the binary random generator. Instead of a single flip-flop, an asynchronous decimal counter (see Section 6.3) is controlled. While S1 is depressed, all digits between 0 and 9 are displayed in rapid succession. One sees all segments, that is, an '8' will be visible. As soon as the button is released, a single random number between 0 and 9 will remain.

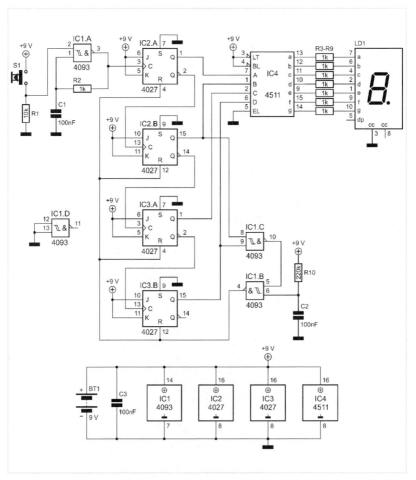

Figure 8.30   *0 – 9 random number generator*

8 APPLICATIONS

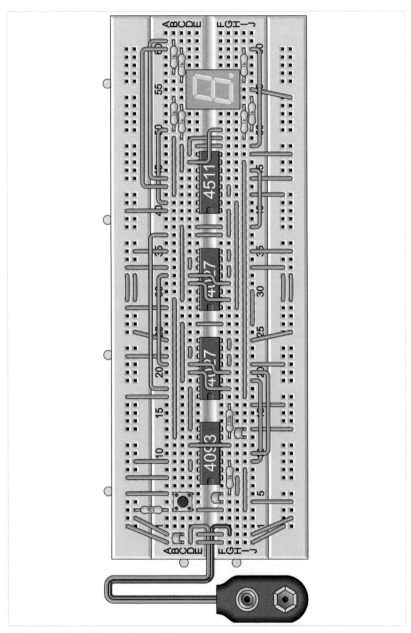

Figure 8.31 *Electronic roulette*

8.15 Digital dice

A dice has six faces, producing values from 1 to 6. Thus, our random number generator requires a six-combination counter.

A three-stage asynchronous counter is used, which, without additional circuitry, will count between 0 and 7. Using an AND operation, we decode the number 7 ('111') on the three Q outputs, which resets the counter, but not to the '0' state: The reset pulse feeds the S input of IC2.A, as well as the R inputs of the two following stages, thus presetting the counter to 1 instead. Thus, our counter works only on numbers between 1 and 6.

We could decode the six possible numbers to derive exactly the six possible die face patterns, but this would add significant complexity. Best to seek some simplifications and link the LEDs directly to the digital counter outputs.

To generate all die patterns, we need seven LEDs arranged in the following manner:

D4		D3
D6	D1	D7
D2		D5

Bit 2	Bit 1	Bit 0	Decimal	D1	D2, D3	D4, D4	D6, D7
0	0	1	1	1	0	0	0
0	1	0	2	0	1	0	0
0	1	1	3	1	1	0	0
1	0	0	4	0	1	1	0
1	0	1	5	1	1	1	0
1	1	0	6	0	1	1	1

Table 8.3   Truth table for electronic dice

8 APPLICATIONS

The middle LED (D1) lights up for all odd numbers. Therefore, it can simply be driven by the Q output on the first flip-flop (Bit 0).

All the other LEDs are connected in pairs. Each pair of LEDs in series requires a small resistor to maintain equal brightness. D2 and D3 light up when Bit 1 or Bit 2 are '1'. D4 and D5 are directly driven by Bit 2. D6 and D7 are used only for the six, and light up when Bit 2 and Bit 1 are '1'.

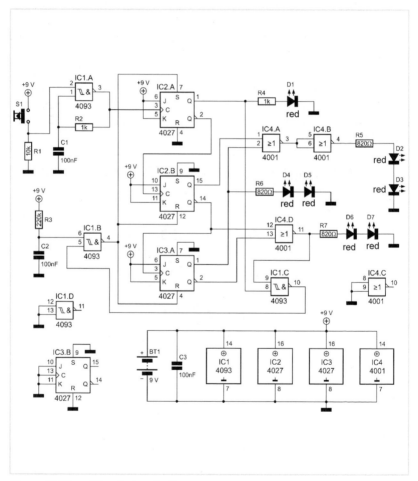

Figure 8.32   The electronic dice

DIGITAL DICE

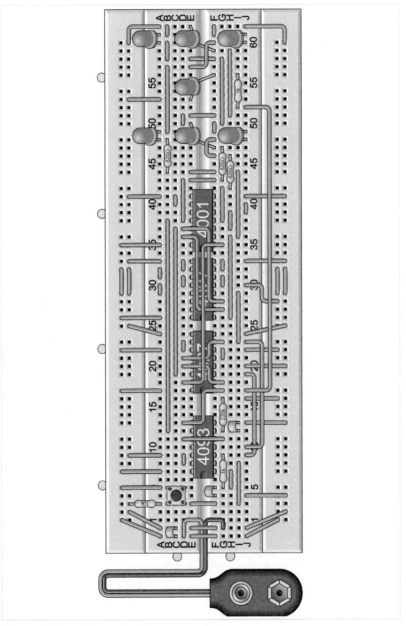

Figure 8.33 *Electronic dice construction*

9 SGS Datasheets

HCF4001B

QUAD 2-INPUT NOR GATE

- PROPAGATION DELAY TIME:
  t_{PD} = 50ns (TYP.) at V_{DD} = 10V C_L = 50pF
- BUFFERED INPUTS AND OUTPUTS
- STANDARDIZED SYMMETRICAL OUTPUT CHARACTERISTICS
- QUIESCENT CURRENT SPECIFIED UP TO 20V
- 5V, 10V AND 15V PARAMETRIC RATINGS
- INPUT LEAKAGE CURRENT
  I_I = 100nA (MAX) AT V_{DD} = 18V T_A = 25°C
- 100% TESTED FOR QUIESCENT CURRENT

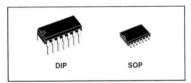

DESCRIPTION
The HCF4001B is a monolithic integrated circuit fabricated in Metal Oxide Semiconductor technology available in DIP and SOP packages. The HCF4001B QUAD 2-INPUT NOR GATE provides the system designer with direct implementation of the NOR function and supplement the existing family of CMOS gates. All inputs and outputs are buffered.

ORDER CODES

PACKAGE	TUBE	T & R
DIP	HCF4001BEY	
SOP	HCF4001BM1	HCF4001M013TR

PIN CONNECTION

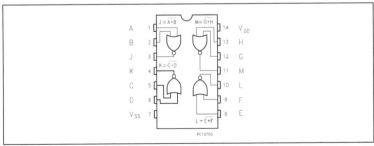

March 2004

HCF4001B

INPUT EQUIVALENT CIRCUIT

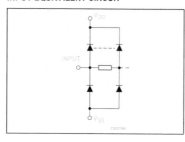

LOGIC DIAGRAM

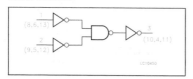

PIN DESCRIPTION

PIN N°	SYMBOL	NAME AND FUNCTION
1, 2, 5, 6, 8, 9, 12, 13	A, B, C, D, E, F, G, H	Data Inputs
3, 4, 10, 11	J, K, L, M	Data Outputs
7	V_{SS}	Negative Supply Voltage
14	V_{DD}	Positive Supply Voltage

TRUTH TABLE

INPUTS		OUTPUTS
A, C, E, G	B, D, F, H	J, K, L, M
L	L	H
L	H	L
H	L	L
H	H	L

ABSOLUTE MAXIMUM RATINGS

Symbol	Parameter	Value	Unit
V_{DD}	Supply Voltage	-0.5 to +22	V
V_I	DC Input Voltage	-0.5 to V_{DD} + 0.5	V
I_I	DC Input Current	± 10	mA
P_D	Power Dissipation per Package	200	mW
	Power Dissipation per Output Transistor	100	mW
T_{op}	Operating Temperature	-55 to +125	°C
T_{stg}	Storage Temperature	-65 to +150	°C

Absolute Maximum Ratings are those values beyond which damage to the device may occur. Functional operation under these conditions is not implied.
All voltage values are referred to V_{SS} pin voltage.

RECOMMENDED OPERATING CONDITIONS

Symbol	Parameter	Value	Unit
V_{DD}	Supply Voltage	3 to 20	V
V_I	Input Voltage	0 to V_{DD}	V
T_{op}	Operating Temperature	-55 to 125	°C

HCF4001B

DC SPECIFICATIONS

Symbol	Parameter	Test Condition				Value						Unit
		V_I (V)	V_O (V)	$|I_O|$ (μA)	V_{DD} (V)	$T_A = 25°C$			-40 to 85°C		-55 to 125°C	
						Min.	Typ.	Max.	Min.	Max.	Min.	Max.	
I_L	Quiescent Current	0/5			5		0.01	0.25		7.5		7.5	μA
		0/10			10		0.01	0.5		15		15	
		0/15			15		0.01	1		30		30	
		0/20			20		0.02	5		150		150	
V_{OH}	High Level Output Voltage	0/5		<1	5	4.95			4.95		4.95		V
		0/10		<1	10	9.95			9.95		9.95		
		0/15		<1	15	14.95			14.95		14.95		
V_{OL}	Low Level Output Voltage	5/0		<1	5			0.05		0.05		0.05	V
		10/0		<1	10			0.05		0.05		0.05	
		15/0		<1	15			0.05		0.05		0.05	
V_{IH}	High Level Input Voltage		0.5/4.5	<1	5	3.5			3.5		3.5		V
			1/9	<1	10	7			7		7		
			1.5/13.5	<1	15	11			11		11		
V_{IL}	Low Level Input Voltage		4.5/0.5	<1	5			1.5		1.5		1.5	V
			9/1	<1	10			3		3		3	
			13.5/1.5	<1	15			4		4		4	
I_{OH}	Output Drive Current	0/5	2.5	<1	5	-1.36	-3.2		-1.15		-1.1		mA
		0/5	4.6	<1	5	-0.44	-1		-0.36		-0.36		
		0/10	9.5	<1	10	-1.1	-2.6		-0.9		-0.9		
		0/15	13.5	<1	15	-3.0	-6.8		-2.4		-2.4		
I_{OL}	Output Sink Current	0/5	0.4	<1	5	0.44	1		0.36		0.36		mA
		0/10	0.5	<1	10	1.1	2.6		0.9		0.9		
		0/15	1.5	<1	15	3.0	6.8		2.4		2.4		
I_I	Input Leakage Current	0/18	Any Input		18		$\pm 10^{-5}$	± 0.1		± 1		± 1	μA
C_I	Input Capacitance		Any Input				5	7.5					pF

The Noise Margin for both "1" and "0" level is: 1V min. with V_{DD}=5V, 2V min. with V_{DD}=10V, 2.5V min. with V_{DD}=15V

DYNAMIC ELECTRICAL CHARACTERISTICS (T_{amb} = 25°C, C_L = 50pF, R_L = 200KΩ, t_r = t_f = 20 ns)

Symbol	Parameter	Test Condition	Value (*)			Unit
		V_{DD} (V)	Min.	Typ.	Max.	
t_{TLH} t_{THL}	Output Transition Time	5		125	250	ns
		10		60	120	
		15		45	90	
t_{PLH} t_{PHL}	Propagation Delay Time	5		100	200	ns
		10		50	100	
		15		40	80	

(*) Typical temperature coefficient for all V_{DD} value is 0.3%/°C.

9 SGS DATASHEETS

HCF4001B

TEST CIRCUIT

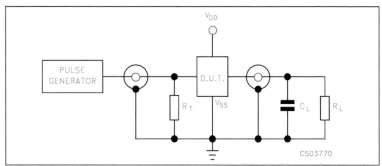

C_L = 50pF or equivalent (includes jig and probe capacitance)
R_L = 200KΩ
R_T = Z_{OUT} of pulse generator (typically 50Ω)

WAVEFORM: PROPAGATION DELAY TIMES (f=1MHz; 50% duty cycle)

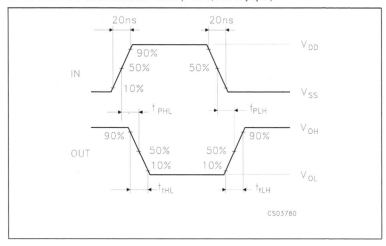

HCF4001B

Plastic DIP-14 MECHANICAL DATA

DIM.	mm.			inch		
	MIN.	TYP	MAX.	MIN.	TYP.	MAX.
a1	0.51			0.020		
B	1.39		1.65	0.055		0.065
b		0.5			0.020	
b1		0.25			0.010	
D			20			0.787
E		8.5			0.335	
e		2.54			0.100	
e3		15.24			0.600	
F			7.1			0.280
I			5.1			0.201
L		3.3			0.130	
Z	1.27		2.54	0.050		0.100

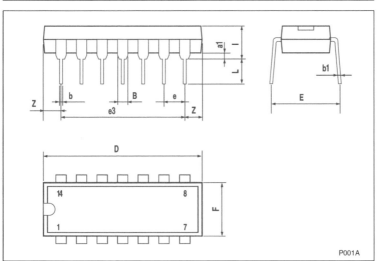

P001A

HCF4001B

SO-14 MECHANICAL DATA

DIM.	mm.			inch		
	MIN.	TYP.	MAX.	MIN.	TYP.	MAX.
A			1.75			0.068
a1	0.1		0.2	0.003		0.007
a2			1.65			0.064
b	0.35		0.46	0.013		0.018
b1	0.19		0.25	0.007		0.010
C		0.5			0.019	
c1			45° (typ.)			
D	8.55		8.75	0.336		0.344
E	5.8		6.2	0.228		0.244
e		1.27			0.050	
e3		7.62			0.300	
F	3.8		4.0	0.149		0.157
G	4.6		5.3	0.181		0.208
L	0.5		1.27	0.019		0.050
M			0.68			0.026
S			8° (max.)			

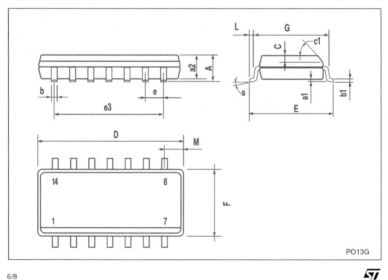

PO13G

HCF4001B

Tape & Reel SO-14 MECHANICAL DATA

DIM.	mm.			inch		
	MIN.	TYP	MAX.	MIN.	TYP.	MAX.
A			330			12.992
C	12.8		13.2	0.504		0.519
D	20.2			0.795		
N	60			2.362		
T			22.4			0.882
Ao	6.4		6.6	0.252		0.260
Bo	9		9.2	0.354		0.362
Ko	2.1		2.3	0.082		0.090
Po	3.9		4.1	0.153		0.161
P	7.9		8.1	0.311		0.319

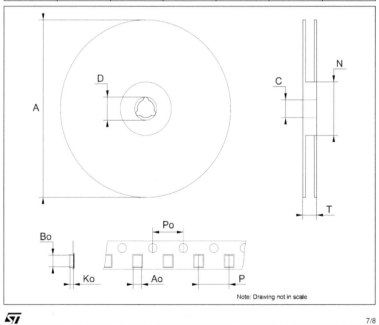

Note: Drawing not in scale

9 SGS DATASHEETS

HCF4027B

DUAL J-K MASTER SLAVE FLIP-FLOP

- SET RESET CAPABILITY
- STATIC FLIP-FLOP OPERATION - RETAINS STATE INDEFINETELY WITH CLOCK LEVEL EITHER "HIGH" OR "LOW"
- MEDIUM-SPEED OPERATION - 16MHz (Typ. clock toggle rate at 10V)
- QUIESCENT CURRENT SPECIFIED UP TO 20V
- STANDARDIZED SYMMETRICAL OUTPUT CHARACTERISTICS
- 5V, 10V AND 15V PARAMETRIC RATINGS
- INPUT LEAKAGE CURRENT
  I_I = 100nA (MAX) AT V_{DD} = 18V T_A = 25°C
- 100% TESTED FOR QUIESCENT CURRENT
- MEETS ALL REQUIREMENTS OF JEDEC JESD13B " STANDARD SPECIFICATIONS FOR DESCRIPTION OF B SERIES CMOS DEVICES"

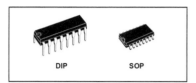

ORDER CODES

PACKAGE	TUBE	T & R
DIP	HCF4027BEY	
SOP	HCF4027BM1	HCF4027M013TR

DESCRIPTION

HCF4027B is a monolithic integrated circuit fabricated in Metal Oxide Semiconductor technology available in DIP and SOP packages.
HCF4027B is a single monolithic chip integrated circuit containing two identical complementary-symmetry J-K master-slave flip-flops. Each flip-flop has provisions for individual J, K, Set, Reset, and Clock input signals. Buffered Q and $\overline{Q}$ signals are provided as outputs. This input-output arrangement provides for compatible operation with the HCF4013B dual D type flip-flop.
This device is useful in performing control, register, and toggle functions. Logic levels present at the J and K inputs, along with internal self-steering, control the state of each flip-flop; changes in the flip-flop state are synchronous with the positive-going transition of the clock pulse. Set and Reset functions are independent of the clock and are initiated when a high level signal is present at either the Set or Reset input.

PIN CONNECTION

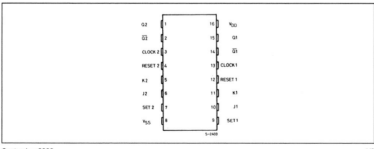

September 2002

HCF4027B

INPUT EQUIVALENT CIRCUIT

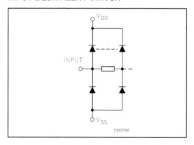

PIN DESCRIPTION

PIN No	SYMBOL	NAME AND FUNCTION
6, 5	J2, K2	Inputs
10, 11	J1, K1	inputs
13, 3	CLOCK1, CLOCK2	Clock Inputs
12, 4	RESET1, RESET2	Reset Inputs
9, 7	SET1, SET2	Set Inputs
1, 2	Q2, $\overline{Q2}$	Outputs
15, 14	Q1, $\overline{Q1}$	Outputs
8	V_{SS}	Negative Supply Voltage
16	V_{DD}	Positive Supply Voltage

FUNCTIONAL DIAGRAM

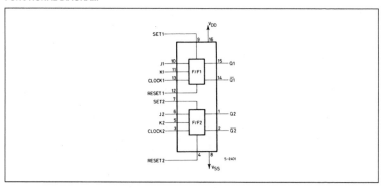

TRUTH TABLE

PRESENT STATE					CLOCK*	NEXT STATE		
Inputs				Output		Outputs		
J	K	S	R	Q		Q	$\overline{Q}$	
H	X	L	L	L	⌐	H	L	
X	L	L	L	H	⌐	H	L	
L	X	L	L	L	⌐	L	H	
X	H	L	L	H	⌐	L	H	
X	X	L	L	X	⌐			NO CHANGE
X	X	H	L	X	X	H	L	
X	X	L	H	X	X	L	H	
X	X	H	H	X	X	H	H	

X : Don"t Care
* : Level Change

HCF4027B

LOGIC DIAGRAM

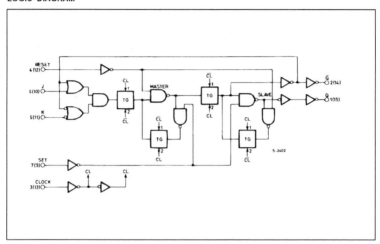

ABSOLUTE MAXIMUM RATINGS

Symbol	Parameter	Value	Unit
V_{DD}	Supply Voltage	-0.5 to +22	V
V_I	DC Input Voltage	-0.5 to V_{DD} + 0.5	V
I_I	DC Input Current	± 10	mA
P_D	Power Dissipation per Package	200	mW
	Power Dissipation per Output Transistor	100	mW
T_{op}	Operating Temperature	-55 to +125	°C
T_{stg}	Storage Temperature	-65 to +150	°C

Absolute Maximum Ratings are those values beyond which damage to the device may occur. Functional operation under these conditions is not implied.
All voltage values are referred to V_{SS} pin voltage.

RECOMMENDED OPERATING CONDITIONS

Symbol	Parameter	Value	Unit
V_{DD}	Supply Voltage	3 to 20	V
V_I	Input Voltage	0 to V_{DD}	V
T_{op}	Operating Temperature	-55 to 125	°C

9 SGS DATASHEETS

HCF4027B

DC SPECIFICATIONS

Symbol	Parameter	Test Condition				Value						Unit
		V_I (V)	V_O (V)	$|I_O|$ (µA)	V_{DD} (V)	$T_A = 25°C$			-40 to 85°C		-55 to 125°C	
						Min.	Typ.	Max.	Min.	Max.	Min.	Max.	
I_L	Quiescent Current	0/5			5		0.02	1		30		30	µA
		0/10			10		0.02	2		60		60	
		0/15			15		0.02	4		120		120	
		0/20			20		0.04	20		600		600	
V_{OH}	High Level Output Voltage	0/5		<1	5	4.95			4.95		4.95		V
		0/10		<1	10	9.95			9.95		9.95		
		0/15		<1	15	14.95			14.95		14.95		
V_{OL}	Low Level Output Voltage	5/0		<1	5		0.05			0.05		0.05	V
		10/0		<1	10		0.05			0.05		0.05	
		15/0		<1	15		0.05			0.05		0.05	
V_{IH}	High Level Input Voltage		0.5/4.5	<1	5	3.5			3.5		3.5		V
			1/9	<1	10	7			7		7		
			1.5/13.5	<1	15	11			11		11		
V_{IL}	Low Level Input Voltage		4.5/0.5	<1	5			1.5		1.5		1.5	V
			9/1	<1	10			3		3		3	
			13.5/1.5	<1	15			4		4		4	
I_{OH}	Output Drive Current	0/5	2.5	<1	5	-1.36	-3.2		-1.15		-1.1		mA
		0/5	4.6	<1	5	-0.44	-1		-0.36		-0.36		
		0/10	9.5	<1	10	-1.1	-2.6		-0.9		-0.9		
		0/15	13.5	<1	15	-3.0	-6.8		-2.4		-2.4		
I_{OL}	Output Sink Current	0/5	0.4	<1	5	0.44	1		0.36		0.36		mA
		0/10	0.5	<1	10	1.1	2.6		0.9		0.9		
		0/15	1.5	<1	15	3.0	6.8		2.4		2.4		
I_I	Input Leakage Current	0/18	Any Input	18			$\pm 10^{-5}$	± 0.1		± 1		± 1	µA
C_I	Input Capacitance		Any Input				5	7.5					pF

The Noise Margin for both "1" and "0" level is: 1V min. with V_{DD}=5V, 2V min. with V_{DD}=10V, 2.5V min. with V_{DD}=15V

HCF4027B

DYNAMIC ELECTRICAL CHARACTERISTICS (T_{amb} = 25°C, C_L = 50pF, R_L = 200KΩ, t_r = t_f = 20 ns)

Symbol	Parameter	Test Condition V_{DD} (V)	Value (*) Min.	Value (*) Typ.	Value (*) Max.	Unit
t_{PLH} t_{PHL}	Propagation Delay Time (Clock to Q or $\overline{Q}$ Outputs)	5		150	300	ns
		10		65	130	
		15		45	90	
t_{PLH}	Propagation Delay Time (Set to Q or Reset to $\overline{Q}$)	5		150	300	ns
		10		65	130	
		15		45	90	
t_{PHL}	Propagation Delay Time (Set to $\overline{Q}$ or Reset to Q)	5		200	400	ns
		10		85	170	
		15		60	120	
t_{TLH} t_{THL}	Transition Time	5		100	200	ns
		10		50	100	
		15		40	80	
t_W	Pulse Width (Clock)	5	140	70		ns
		10	60	30		
		15	40	20		
t_W	Pulse Width (Set or Reset)	5	180	90		ns
		10	80	40		
		15	50	25		
t_r, t_f	Clock input Rise or Fall Time	5			15	μs
		10			4	
		15			1	
t_{setup}	Setup Time (DATA)	5	200	100		ns
		10	75	35		
		15	50	25		
f_{MAX}	Maximum Clock Input Frequency [1] (toggle mode)	5	3.5	7		MHz
		10	8	16		
		15	12	24		

(*) Typical temperature coefficient for all V_{DD} value is 0.3 %/°C.
(1) Input t_r, t_f = 5ns

HCF4027B

TEST CIRCUIT

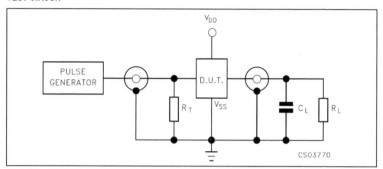

C_L = 50pF or equivalent (includes jig and probe capacitance)
R_L = 200KΩ
R_T = Z_{OUT} of pulse generator (typically 50Ω)

WAVEFORM : PROPAGATION DELAY TIMES, MINIMUM PULSE WIDTH (CK), SETUP AND HOLD TIME (J or K to CK) (f=1MHz; 50% duty cycle)

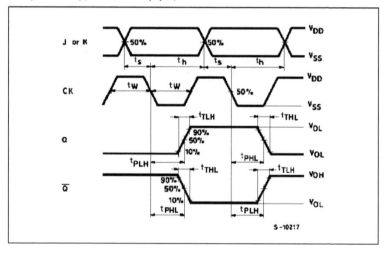

HCF4027B

Plastic DIP-16 (0.25) MECHANICAL DATA

DIM.	mm.			inch		
	MIN.	TYP	MAX.	MIN.	TYP.	MAX.
a1	0.51			0.020		
B	0.77		1.65	0.030		0.065
b		0.5			0.020	
b1		0.25			0.010	
D			20			0.787
E		8.5			0.335	
e		2.54			0.100	
e3		17.78			0.700	
F			7.1			0.280
I			5.1			0.201
L		3.3			0.130	
Z			1.27			0.050

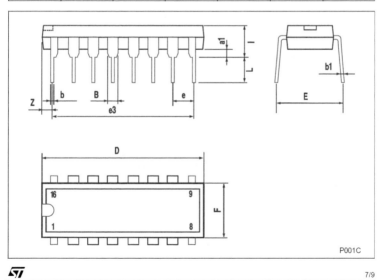

P001C

HCF4027B

SO-16 MECHANICAL DATA

DIM.	mm.			inch		
	MIN.	TYP	MAX.	MIN.	TYP.	MAX.
A			1.75			0.068
a1	0.1		0.2	0.003		0.007
a2			1.65			0.064
b	0.35		0.46	0.013		0.018
b1	0.19		0.25	0.007		0.010
C		0.5			0.019	
c1			45° (typ.)			
D	9.8		10	0.385		0.393
E	5.8		6.2	0.228		0.244
e		1.27			0.050	
e3		8.89			0.350	
F	3.8		4.0	0.149		0.157
G	4.6		5.3	0.181		0.208
L	0.5		1.27	0.019		0.050
M			0.62			0.024
S			8° (max.)			

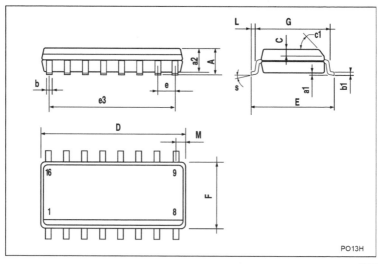

PO13H

HCF4093B

QUAD 2 INPUT NAND SCHMITT TRIGGER

- SCHMITT TRIGGER ACTION ON EACH INPUT WITH NO EXTERNAL COMPONENTS
- HYSTERESIS VOLTAGE TYPICALLY 0.9V at V_{DD} = 5V AND 2.3V at V_{DD} = 10V
- NOISE IMMUNITY GREATER THAN 50% OF V_{DD} (Typ.)
- NO LIMIT ON INPUT RISE AND FALL TIMES
- QUIESCENT CURRENT SPECIFIED UP TO 20V
- STANDARDIZED SYMMETRICAL OUTPUT CHARACTERISTICS
- 5V, 10V AND 15V PARAMETRIC RATINGS
- INPUT LEAKAGE CURRENT I_I = 100nA (MAX) AT V_{DD} = 18V T_A = 25°C
- 100% TESTED FOR QUIESCENT CURRENT
- MEETS ALL REQUIREMENTS OF JEDEC JESD13B " STANDARD SPECIFICATIONS FOR DESCRIPTION OF B SERIES CMOS DEVICES"

DESCRIPTION
The HCF4093B is a monolithic integrated circuit fabricated in Metal Oxide Semiconductor technology available in DIP and SOP packages.

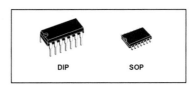

ORDER CODES

PACKAGE	TUBE	T & R
DIP	HCF4093BEY	
SOP	HCF4093BM1	HCF4093M013TR

The HCF4093B type consists of four schmitt trigger circuits. Each circuit functions as a two input NAND gate with schmitt trigger action on both inputs. The gate switches at different points for positive and negative going signals. The difference between the positive voltage (V_P) and the negative voltage (V_N) is defined as hysteresis voltage (V_H).

PIN CONNECTION

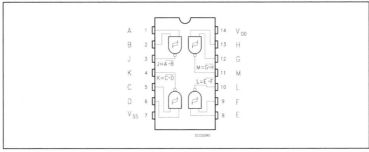

September 2001

HCF4093B

INPUT EQUIVALENT CIRCUIT

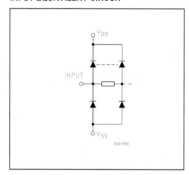

PIN DESCRIPTION

PIN No	SYMBOL	NAME AND FUNCTION
1, 2, 5, 6, 8, 9, 12, 13	A, B, C, D, E, F, G, H	Data Inputs
3, 4, 10, 11	J, K, L, M	Data Outputs
7	V_{SS}	Negative Supply Voltage
14	V_{DD}	Positive Supply Voltage

TRUTH TABLE

INPUTS		OUTPUTS
A, C, E, G	B, D, F, H	J, K, L, M
L	L	H
L	H	H
H	L	H
H	H	L

ABSOLUTE MAXIMUM RATINGS

Symbol	Parameter	Value	Unit
V_{DD}	Supply Voltage	-0.5 to +22	V
V_I	DC Input Voltage	-0.5 to V_{DD} + 0.5	V
I_I	DC Input Current	± 10	mA
P_D	Power Dissipation per Package	200	mW
	Power Dissipation per Output Transistor	100	mW
T_{op}	Operating Temperature	-55 to +125	°C
T_{stg}	Storage Temperature	-65 to +150	°C

Absolute Maximum Ratings are those values beyond which damage to the device may occur. Functional operation under these conditions is not implied.
All voltage values are referred to V_{SS} pin voltage.

RECOMMENDED OPERATING CONDITIONS

Symbol	Parameter	Value	Unit
V_{DD}	Supply Voltage	3 to 20	V
V_I	Input Voltage	0 to V_{DD}	V
T_{op}	Operating Temperature	-55 to 125	°C

HCF4093B

DC SPECIFICATIONS

Symbol	Parameter	Test Condition				Value						Unit
		V_I (V)	V_O (V)	$|I_O|$ (μA)	V_{DD} (V)	$T_A = 25°C$			-40 to 85°C		-55 to 125°C	
						Min.	Typ.	Max.	Min.	Max.	Min.	Max.	
I_L	Quiescent Current	0/5			5		0.02	1		30		30	μA
		0/10			10		0.02	2		60		60	
		0/15			15		0.02	4		120		120	
		0/20			20		0.04	20		600		600	
V_{OH}	High Level Output Voltage	0/5		<1	5	4.95			4.95		4.95		V
		0/10		<1	10	9.95			9.95		9.95		
		0/15		<1	15	14.95			14.95		14.95		
V_{OL}	Low Level Output Voltage	5/0		<1	5			0.05		0.05		0.05	V
		10/0		<1	10			0.05		0.05		0.05	
		15/0		<1	15			0.05		0.05		0.05	
V_P	Positive Trigger Threshold Voltage	a			5	2.2	2.9	3.6	2.2	3.6	2.2	3.6	V
		a			10	4.6	5.9	7.1	4.6	7.1	4.6	7.1	
		a			15	6.8	8.8	10.8	6.8	10.8	6.8	10.8	
		b			5	2.6	3.3	4.0	2.6	4	2.6	4	
		b			10	5.6	7	8.2	5.6	8.2	5.6	8.2	
		b			15	6.3	9.4	12.7	6.3	12.7	6.3	12.7	
V_N	Negative Trigger Threshold Voltage	a			5	0.9	1.9	2.8	0.9	2.8	0.9	2.8	V
		a			10	2.5	3.9	5.2	2.5	5.2	2.5	5.2	
		a			15	4	5.8	7.4	4	7.4	4	7.4	
		b			5	1.4	2.3	3.2	1.4	3.2	1.4	3.2	
		b			10	3.4	5.1	6.6	3.4	6.6	3.4	6.6	
		b			15	4.8	7.3	9.6	4.8	9.6	4.8	9.6	
V_H	Hysteresis Voltage	a			5	0.3	0.9	1.6	0.3	1.6	0.3	1.6	V
		a			10	1.2	2.3	3.4	1.2	3.4	1.2	3.4	
		a			15	1.6	3.5	5	1.6	5	1.6	5	
		b			5	0.3	0.9	1.6	0.3	1.6	0.3	1.6	
		b			10	1.2	2.3	3.4	1.2	3.4	1.2	3.4	
		b			15	1.6	3.5	5	1.6	5	1.6	5	
I_{OH}	Output Drive Current	0/5	2.5	<1	5	-1.36	-3.2		-1.15		-1.1		mA
		0/5	4.6	<1	5	-0.44	-1		-0.36		-0.36		
		0/10	9.5	<1	10	-1.1	-2.6		-0.9		-0.9		
		0/15	13.5	<1	15	-3.0	-6.8		-2.4		-2.4		
I_{OL}	Output Sink Current	0/5	0.4	<1	5	0.44	1		0.36		0.36		mA
		0/10	0.5	<1	10	1.1	2.6		0.9		0.9		
		0/15	1.5	<1	15	3.0	6.8		2.4		2.4		
I_I	Input Leakage Current	0/18	Any Input		18		$\pm 10^{-5}$	± 0.1		± 1		± 1	μA
C_I	Input Capacitance		Any Input				5	7.5					pF

The Noise Margin for both "1" and "0" level is: 1V min. with V_{DD}=5V, 2V min. with V_{DD}=10V, 2.5V min. with V_{DD}=15V
a : Input on terminals 1, 5, 8, 12 or 2, 6, 9, 13; other inputs to V_{DD}.
b : Input on terminals 1 and 2, 5 and 6, 8 and 9, or 12 and 13; other inputs to V_{DD}.

9 SGS DATASHEETS

HCF4093B

DYNAMIC ELECTRICAL CHARACTERISTICS ($T_{amb} = 25°C$, $C_L = 50pF$, $R_L = 200K\Omega$, $t_r = t_f = 20$ ns)

Symbol	Parameter	Test Condition		Value (*)		Unit
		V_{DD} (V)	Min.	Typ.	Max.	
t_{PLH} t_{PHL}	Propagation Delay Time	5		190	380	
		10		90	180	ns
		15		65	130	
t_{TLH} t_{THL}	Output Transition Time	5		100	200	
		10		50	100	ns
		15		40	80	

(*) Typical temperature coefficient for all V_{DD} value is 0.3 %/°C.

TEST CIRCUIT

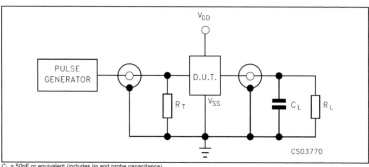

C_L = 50pF or equivalent (includes jig and probe capacitance)
R_L = 200KΩ
R_T = Z_{OUT} of pulse generator (typically 50Ω)

WAVEFORM : PROPAGATION DELAY TIMES (f=1MHz; 50% duty cycle)

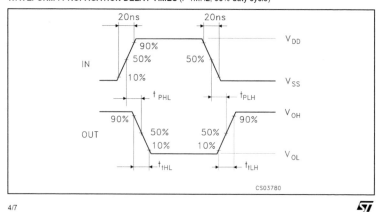

HCF4093B

Plastic DIP-14 MECHANICAL DATA

DIM.	mm.			inch		
	MIN.	TYP.	MAX.	MIN.	TYP.	MAX.
a1	0.51			0.020		
B	1.39		1.65	0.055		0.065
b		0.5			0.020	
b1		0.25			0.010	
D			20			0.787
E		8.5			0.335	
e		2.54			0.100	
e3		15.24			0.600	
F			7.1			0.280
I			5.1			0.201
L		3.3			0.130	
Z	1.27		2.54	0.050		0.100

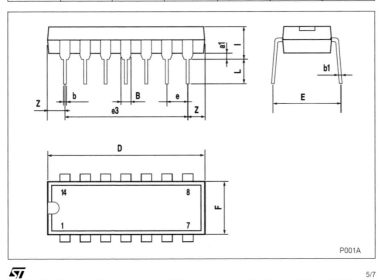

P001A

HCF4093B

SO-14 MECHANICAL DATA

DIM.	mm. MIN.	mm. TYP	mm. MAX.	inch MIN.	inch TYP.	inch MAX.
A			1.75			0.068
a1	0.1		0.2	0.003		0.007
a2			1.65			0.064
b	0.35		0.46	0.013		0.018
b1	0.19		0.25	0.007		0.010
C		0.5			0.019	
c1	45° (typ.)					
D	8.55		8.75	0.336		0.344
E	5.8		6.2	0.228		0.244
e		1.27			0.050	
e3		7.62			0.300	
F	3.8		4.0	0.149		0.157
G	4.6		5.3	0.181		0.208
L	0.5		1.27	0.019		0.050
M			0.68			0.026
S	8° (max.)					

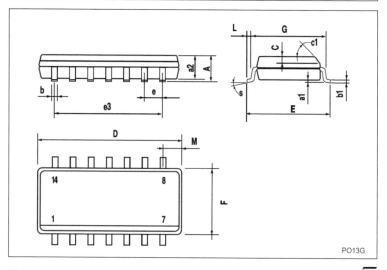

HCF4511B

BCD TO SEVEN SEGMENT LATCH/DECODER/DRIVER

- HIGH OUTPUT SOURCING CAPABILITY (up to 25mA).
- INPUT LATCHES FOR BCD CODE STORAGE
- LAMP TEST AND BLANKING CAPABILITY.
- 7-SEGMENT OUTPUTS BLANKED FOR BCD INPUT CODES > 1001
- QUIESCENT CURRENT SPECIF. UP TO 20V
- STANDARDIZED SYMMETRICAL OUTPUT CHARACTERISTICS
- 5V, 10V, AND 15V PARAMETRIC RATINGS
- INPUT LEAKAGE CURRENT
  I_I = 100nA (MAX) AT V_{DD} = 18V T_A = 25°C
- 100% TESTED FOR QUIESCENT CURRENT
- MEETS ALL REQUIREMENTS OF JEDEC JESD13B "STANDARD SPECIFICATIONS FOR DESCRIPTION OF B SERIES CMOS DEVICES"

ORDER CODES

PACKAGE	TUBE	T & R
DIP	HCF4511BEY	
SOP	HCF4511BM1	HCF4511M013TR

DESCRIPTION

HCF4511B is a monolithic integrated circuit fabricated in Metal Oxide Semiconductor technology available in DIP and SOP packages. HCF4511B is a BCD to 7 segment decoder driver made up of CMOS logic and n-p-n bipolar transistor output devices on a single monolithic structure. This device combines the low quiescent power dissipation and high noise immunity features of CMOS with n-p-n bipolar output transistor capable of sourcing up to 25mA. This capability allows HCF4511B to drive LEDs and other displays directly.
Lamp Test ($\overline{LT}$), Blanking ($\overline{BL}$), and Latch Enable or Strobe inputs are provided to test the display, shut off or intensity-modulate it, and store or strobe a BCD code, respectively. Several different signals may be multiplexed and displayed when external multiplexing circuitry is used.

PIN CONNECTION

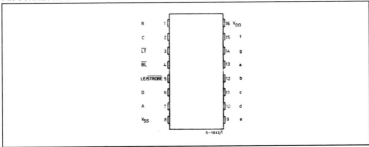

October 2002

HCF4511B

INPUT EQUIVALENT CIRCUIT

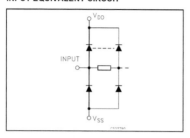

PIN DESCRIPTION

PIN No	SYMBOL	NAME AND FUNCTION
7, 1, 2, 6	A, B, C, D	Bcd Inputs
13, 12, 11, 10, 9, 15, 14	a to g	7-Segment Outputs
3	$\overline{LT}$	Lamp Test Input
4	$\overline{BL}$	Blanking Input
5	LE/$\overline{STROBE}$	Latch Enable or Strobe Input
8	V_{SS}	Negative Supply Voltage
16	V_{DD}	Positive Supply Voltage

FUNCTIONAL DIAGRAM

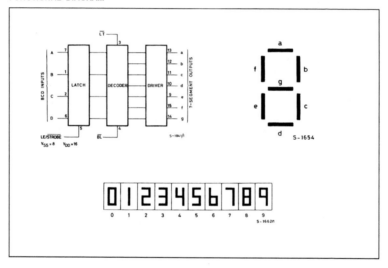

HCF4511B

LOGIC DIAGRAM

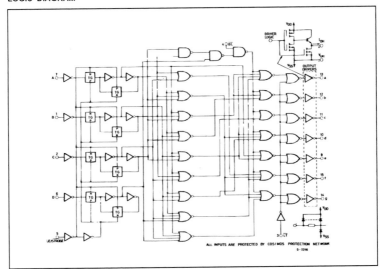

TRUTH TABLE

LE	BL	LT	D	C	B	A	a	b	c	d	e	f	g	DISPLAY
X	X	L	X	X	X	X	H	H	H	H	H	H	H	8
X	L	H	X	X	X	X	L	L	L	L	L	L	L	Blank
L	H	H	L	L	L	L	H	H	H	H	H	H	L	0
L	H	H	L	L	L	H	L	H	H	L	L	L	L	1
L	H	H	L	L	H	L	H	H	L	H	H	L	H	2
L	H	H	L	L	H	H	H	H	H	H	L	L	H	3
L	H	H	L	H	L	L	L	H	H	L	L	H	H	4
L	H	H	L	H	L	H	H	L	H	H	L	H	H	5
L	H	H	L	H	H	L	L	L	H	H	H	H	H	6
L	H	H	L	H	H	H	H	H	H	L	L	L	L	7
L	H	H	H	L	L	L	H	H	H	H	H	H	H	8
L	H	H	H	L	L	H	H	H	H	L	L	H	H	9
L	H	H	H	L	H	L	L	L	L	L	L	L	L	Blank
L	H	H	H	L	H	H	L	L	L	L	L	L	L	Blank
L	H	H	H	H	L	L	L	L	L	L	L	L	L	Blank
L	H	H	H	H	L	H	L	L	L	L	L	L	L	Blank
L	H	H	H	H	H	L	L	L	L	L	L	L	L	Blank
L	H	H	H	H	H	H	L	L	L	L	L	L	L	Blank
H	H	H	X	X	X	X				*				*

X: Don't Care

9 SGS DATASHEETS

HCF4511B

ABSOLUTE MAXIMUM RATINGS

Symbol	Parameter	Value	Unit
V_{DD}	Supply Voltage	-0.5 to +22	V
V_I	DC Input Voltage	-0.5 to V_{DD} + 0.5	V
I_I	DC Input Current	± 10	mA
P_D	Power Dissipation per Package	200	mW
	Power Dissipation per Output Transistor	100	mW
T_{op}	Operating Temperature	-55 to +125	°C
T_{stg}	Storage Temperature	-65 to +150	°C

Absolute Maximum Ratings are those values beyond which damage to the device may occur. Functional operation under these conditions is not implied.

All voltage values are referred to V_{SS} pin voltage.

RECOMMENDED OPERATING CONDITIONS

Symbol	Parameter	Value	Unit
V_{DD}	Supply Voltage	3 to 20	V
V_I	Input Voltage	0 to V_{DD}	V
T_{op}	Operating Temperature	-55 to 125	°C

HCF4511B

DC SPECIFICATIONS

Symbol	Parameter	Test Condition				Value						Unit
		V_I (V)	V_O (V)	$\|I_O\|$ (μA)	V_{DD} (V)	$T_A = 25°C$			-40 to 85°C		-55 to 125°C	
						Min.	Typ.	Max.	Min.	Max.	Min.	Max.	
I_L	Quiescent Current	0/5			5		0.04	5		150		150	μA
		0/10			10		0.04	10		300		300	
		0/15			15		0.04	20		600		600	
		0/20			20		0.08	100		3000		3000	
V_{OH}	High Level Output Voltage	0/5			5	4.95			4.95		4.95		V
		0/10			10	9.95			9.95		9.95		
		0/15			15	14.95			14.95		14.95		
V_{OL}	Low Level Output Voltage	5/0			5			0.05		0.05		0.05	V
		10/0			10			0.05		0.05		0.05	
		15/0			15			0.05		0.05		0.05	
V_{IH}	High Level Input Voltage		0.5/3.8		5	3.5			3.5		3.5		V
			1/8.8		10	7			7		7		
			1.5/13.8		15	11			11		11		
V_{IL}	Low Level Input Voltage		3.8/0.5		5			1.5		1.5		1.5	V
			8.8/1		10			3		3		3	
			13.8/1.5		15			4		4		4	
V_{OH}	Output Drive Voltage			0	5	4.1	4.57		4.1		4.1		V
				5			4.24						
				10		3.6	4.12		3.3		3.3		
				15			3.94						
				20		2.8	3.75		2.5		2.5		
				25			3.54						
				0	10	9.1	9.58		9.1		9.1		V
				5			9.26						
				10		8.75	9.17		8.45		8.45		
				15			9.04						
				20		8.1	8.90		7.8		7.8		
				25			8.75						
				0	15	14.1	14.59		14.1		14.1		V
				5			14.27						
				10		13.75	14.18		13.45		13.45		
				15			14.07						
				20		13.1	13.95		12.8		12.8		
				25			13.80						
I_{OL}	Output Sink Current	0/5	0.4		5	0.44	1		0.36		0.36		mA
		0/10	0.5		10	1.1	2.6		0.9		0.9		
		0/15	1.5		15	3	6.8		2.4		2.4		
I_I	Input Leakage Current (any input)	0/18			18		$\pm 10^{-5}$	± 0.1		± 1		± 1	μA
C_I	Input Capacitance (any input)						5	7.5					pF

The Noise Margin for both "1" and "0" level is: 1V min. with V_{DD}=5V, 2V min. with V_{DD}=10V, 2.5V min. with V_{DD}=15V

HCF4511B

DYNAMIC ELECTRICAL CHARACTERISTICS (T_{amb} = 25°C, C_L = 50pF, R_L = 200KΩ, $t_r = t_f$ = 20 ns)

Symbol	Parameter	V_{DD} (V)	TEST CONDITION	Min.	Typ.	Max.	Unit
t_{PHL}	Propagation Delay Time (DATA)	5			520	1040	
		10			210	420	ns
		15			150	300	
t_{PLH}	Propagation Delay Time (DATA)	5			660	1320	
		10			260	520	ns
		15			180	360	
t_{PHL}	Propagation Delay Time (BL)	5			350	700	
		10			175	350	ns
		15			125	250	
t_{PLH}	Propagation Delay Time (BL)	5			400	800	
		10			175	350	ns
		15			150	300	
t_{PHL}	Propagation Delay Time (LT)	5			250	500	
		10			125	250	ns
		15			85	170	
t_{PLH}	Propagation Delay Time (LT)	5			150	300	
		10			75	150	ns
		15			50	100	
t_{TLH}	Transition Time	5			40	80	
		10			30	60	ns
		15			20	50	
t_{THL}	Transition Time	5			125	310	
		10			75	185	ns
		15			65	160	
t_{setup}	Setup Time	5		150	75		
		10		70	35		ns
		15		40	20		
t_{hold}	Hold Time	5		0	-75		
		10		0	-35		ns
		15		0	-20		
t_W	Strobe Pulse Width	5		400	200		
		10		160	80		ns
		15		100	50		

(*) Typical temperature coefficient for all V_{DD} value is 0.3 %/°C.

HCF4511B

TYPICAL APPLICATIONS (Interfacing with various displays)

Driving Common-cathode 7 Segment Led Displays

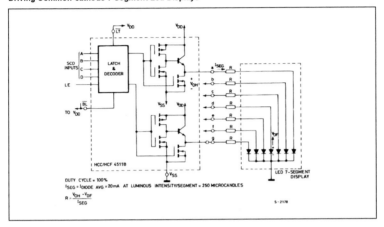

DUTY CYCLE = 100%
$I_{SEG} = I_{DIODE}$ AVG = 20mA AT LUMINOUS INTENSITY/SEGMENT = 250 MICROCANDLES
$R = \dfrac{V_{OH} - V_{DF}}{I_{SEG}}$

Driving Low-voltage Fluorescent Displays

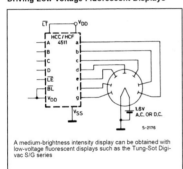

A medium-brightness intensity display can be obtained with low-voltage fluorescent displays such as the Tung-Sot Digivac S/G series

Driving Incandescent Displays

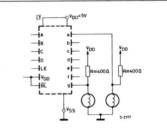

2 Of 7 Segment Shown Connected
Resistor R from VDD VDD to each 7-segment driver output are chosen to keep all Numitron segments slightly on and warm

HCF4511B

TEST CIRCUIT

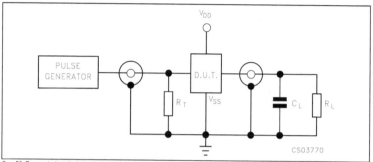

C_L = 50pF or equivalent (includes jig and probe capacitance)
R_L = 200KΩ
R_T = Z_{OUT} of pulse generator (typically 50Ω)

WAVEFORM 1 : PROPAGATION DELAY TIMES (f=1MHz; 50% duty cycle)

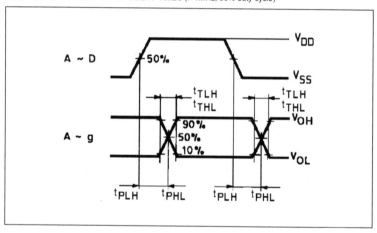

HCF4511B

WAVEFORM 2 : MINIMUM PULSE WIDTH (f=1MHz; 50% duty cycle)

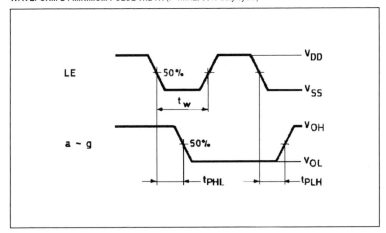

WAVEFORM 3 : PROPAGATION DELAY TIMES (f=1MHz; 50% duty cycle)

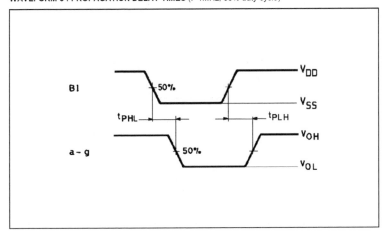

HCF4511B

WAVEFORM 4 : PROPAGATION DELAY TIMES (f=1MHz; 50% duty cycle)

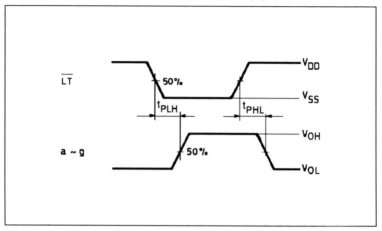

WAVEFORM 5 : MINIMUM SETUP AND HOLD TIME (f=1MHz; 50% duty cycle)

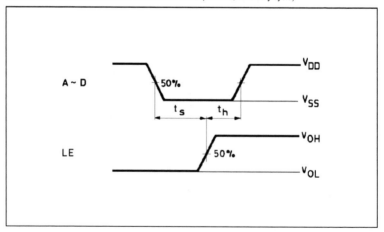

HCF4511B

Plastic DIP-16 (0.25) MECHANICAL DATA

DIM.	mm.			inch		
	MIN.	TYP	MAX.	MIN.	TYP.	MAX.
a1	0.51			0.020		
B	0.77		1.65	0.030		0.065
b		0.5			0.020	
b1		0.25			0.010	
D			20			0.787
E		8.5			0.335	
e		2.54			0.100	
e3		17.78			0.700	
F			7.1			0.280
I			5.1			0.201
L		3.3			0.130	
Z			1.27			0.050

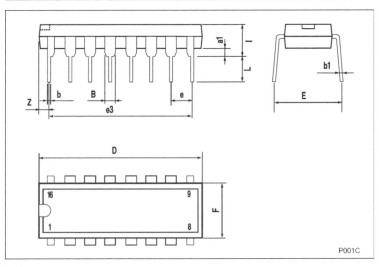

P001C

9 SGS DATASHEETS

HCF4511B

SO-16 MECHANICAL DATA

DIM.	mm.			inch		
	MIN.	TYP	MAX.	MIN.	TYP.	MAX.
A			1.75			0.068
a1	0.1		0.2	0.003		0.007
a2			1.65			0.064
b	0.35		0.46	0.013		0.018
b1	0.19		0.25	0.007		0.010
C		0.5			0.019	
c1			45° (typ.)			
D	9.8		10	0.385		0.393
E	5.8		6.2	0.228		0.244
e		1.27			0.050	
e3		8.89			0.350	
F	3.8		4.0	0.149		0.157
G	4.6		5.3	0.181		0.208
L	0.5		1.27	0.019		0.050
M			0.62			0.024
S			8° (max.)			

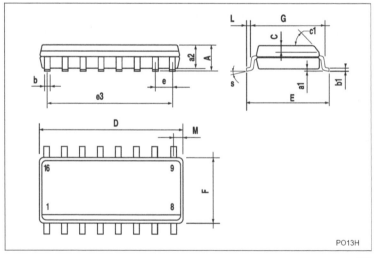

PO13H

HCF4511B

Information furnished is believed to be accurate and reliable. However, STMicroelectronics assumes no responsibility for the consequences of use of such information nor for any infringement of patents or other rights of third parties which may result from its use. No license is granted by implication or otherwise under any patent or patent rights of STMicroelectronics. Specifications mentioned in this publication are subject to change without notice. This publication supersedes and replaces all information previously supplied. STMicroelectronics products are not authorized for use as critical components in life support devices or systems without express written approval of STMicroelectronics.

© The ST logo is a registered trademark of STMicroelectronics

© 2002 STMicroelectronics - Printed in Italy - All Rights Reserved
STMicroelectronics GROUP OF COMPANIES
Australia - Brazil - Canada - China - Finland - France - Germany - Hong Kong - India - Israel - Italy - Japan - Malaysia - Malta - Morocco - Singapore - Spain - Sweden - Switzerland - United Kingdom - United States.
© http://www.st.com

Index

A
Alarm system 111
AND function 18, 31, 41
AND gate 21, 31, 41
Antivalence function 49
Asynchronous counter 73
Automatic night light 109

B
Binary number 17
Binary state 17
Bit 17
Blinking light 95
Breadboard 11

C
Circular shift register 70
Clock pulse 62, 65
CMOS 4000 series 11, 20
Contact bounce 62
Counter 73

D
Debounce circuit 62
Dice 135
Digital electronics 11

E
Equivalence function 51
Exclusive-OR 49

F
Flip-flop 55
Flip-flop blinker 97
Forward voltage 24
Four-bit counter 76
Frequency 95

H
Hallway light timer 128

I
Inverter 21, 26

J
JK flip-flop 60, 62

L
LED 24
Light-activated alarm system 113
Light-controlled oscillator 103

M
Majority function 53
Memory 55
Metronome 99
Mini organ 105
Monoflop 126
Morse Code practice device 101

N
NAND function 26, 43

INDEX

NAND gate 22, 23, 29, 43
Nine-to-zero counter 93
NOR gate 33, 35, 47
NOT function 19, 21
NOT gate 26

O
One-bit counter 73
OR function 18, 39
OR gate 33, 39, 45
Oscillator 95
Oscilloscope 26

P
Piezoelectric transducer 99
Pull-down resistor 23

R
Random generator 131
Roulette 133
RS flip-flop 55
RS flip-flop from NAND gates 58
RS flip-flop from NOR gates 55
Running light 115

S
Schmitt trigger 22
Seven-segment decoder 89
Seven-segment display 87

Shift register 65
Siren 107
Supply voltage 24
Synchronous counter 79

T
Time switch 126
Tone generator 101
Traffic light control 119
Troubleshooting 26
Truth table 24
TTL 7400 series 19
Turn-off delay 122
Turn-on delay 124
Twilight switch 109
Two-bit counter 73, 76

U
Up-down counter 82

V
Voltmeter 26

X
XNOR function 51
XOR function 49

Z
Zero-to-nine counter 91